Sustainable Development Goals Series

The **Sustainable Development Goals Series** is Springer Nature's inaugural cross-imprint book series that addresses and supports the United Nations' seventeen Sustainable Development Goals. The series fosters comprehensive research focused on these global targets and endeavours to address some of society's greatest grand challenges. The SDGs are inherently multidisciplinary, and they bring people working across different fields together and working towards a common goal. In this spirit, the Sustainable Development Goals series is the first at Springer Nature to publish books under both the Springer and Palgrave Macmillan imprints, bringing the strengths of our imprints together.

The Sustainable Development Goals Series is organized into eighteen subseries: one subseries based around each of the seventeen respective Sustainable Development Goals, and an eighteenth subseries, "Connecting the Goals," which serves as a home for volumes addressing multiple goals or studying the SDGs as a whole. Each subseries is guided by an expert Subseries Advisor with years or decades of experience studying and addressing core components of their respective Goal.

The SDG Series has a remit as broad as the SDGs themselves, and contributions are welcome from scientists, academics, policymakers, and researchers working in fields related to any of the seventeen goals. If you are interested in contributing a monograph or curated volume to the series, please contact the Publishers: Zachary Romano [Springer; zachary.romano@springer.com] and Rachael Ballard [Palgrave Macmillan; rachael.ballard@palgrave.com].

Fingani Annie Mphande

Sustainable Health in Low and Middle Income Countries

Achieving SDG3 in the (Post) Pandemic World

Springer

Fingani Annie Mphande
Faculty of Medicine
King Mongkut's Institute of Technology Ladkrabang
Bangkok, Thailand

ISSN 2523-3084 ISSN 2523-3092 (electronic)
Sustainable Development Goals Series
ISBN 978-981-99-4253-4 ISBN 978-981-99-4254-1 (eBook)
https://doi.org/10.1007/978-981-99-4254-1

This Springer imprint is published by the registered company Springer Nature Singapore Pte Ltd.
The registered company address is: 152 Beach Road, #21-01/04 Gateway East, Singapore 189721, Singapore

To My Dad (a CU)

Preface

Low- and middle-income countries (LMIC) have struggled with weak healthcare systems, often overwhelmed by the burden of communicable and non-communicable diseases, which are further complicated by emerging and re-emerging diseases. The healthcare systems in LMIC are far from sustainable with most surviving from donor funding and grants for specific health initiatives and programmes. This book is going to unravel some of the issues that have contributed to such healthcare systems and how the lessons from the COVID-19 pandemic could be a steppingstone in building resilient and sustainable healthcare systems in LMIC. The book is intended for healthcare workers, researchers, policy makers, governments, students, non-profit and non-governmental organisations (NGO), and many more who contribute to attaining sustainable health globally.

The essence of United Nations Sustainable Development Goals is in the phrase "To Leave No One Behind". Sustainable Development Goal 3 (SDG3) targets "Good Health and Well-being; Ensure healthy lives and promote well-being for all at all ages". This book will focus on parts of SDG3, specifically SDG3.3 and 3b, 3c and 3d, targets linked to the threat of communicable diseases and their elimination. As populations worldwide face the COVID-19 pandemic, while fighting the HIV epidemic and other emerging and re-emerging infectious diseases SDG3, "Ensure healthy lives and well-being for all at all ages" comes back into focus. Several targets have been set to meet SDG3, this book will focus on these targets.

Target 3.3: By 2030, end the epidemics of AIDS, tuberculosis, malaria, and neglected tropical diseases and combat hepatitis, water-borne diseases, and other communicable diseases.

Target 3.B: Support the research and development of vaccines and medicines for the communicable and noncommunicable diseases that primarily affect developing countries, provide access to affordable essential medicines and vaccines, in accordance with the Doha Declaration on the TRIPS Agreement and Public Health, which affirms the right of developing countries to use to the full the provisions in the Agreement on Trade Related Aspects of Intellectual Property Rights regarding flexibilities to protect public health, and, in particular, provide access to medicines for all.

Target 3.C: Substantially increase health financing and the recruitment, development, training, and retention of the health workforce in developing countries, especially in least developed countries and small island developing States.

Target 3.D: Strengthen the capacity of all countries, in particular developing countries, for early warning, risk reduction, and management of national and global health risks.

Despite strides being made in some areas for target 3.3, the COVID-19 pandemic has caused interruptions that will considerably affect vaccination coverage as well as the progress that was made, for example, in reducing malaria, HIV, and TB cases in several endemic countries. Vulnerable populations, who were already struggling to access their healthcare needs before the pandemic, may face even greater challenges at present and in the years to come post-pandemic. This book will focus on the success and challenges in low- and middle-income countries (LMIC) post-COVID-19 in combating various infectious diseases and the progress in attaining these targets. The overarching question in this book is, can sustainable health be attained LMIC?

Bangkok, Thailand Fingani Annie Mphande

Acknowledgements

I would like to extend my sincere gratitude to my family for their support throughout the writing process. To my colleagues and friends who encouraged me throughout the project. To Alexandra Campbell my editor and all the Springer Nature Team who worked tirelessly with me throughout the publication process. Sincere gratitude to WHO permissions team for their help and swift response and for granting permission to use the figures from WHO sources.

Contents

Abbreviations

ART	Antiretroviral Treatment
CCHF	Crimean-Congo haemorrhagic fever
CCHFV	Crimean-Congo haemorrhagic fever virus
CDC	Centres for Disease Control
COVID-19	Coronavirus disease of 2019
EBV	Ebola virus disease
ECDC	European Centres for Disease Control
eHL	Electronic Health Literacy
FCAS	Fragile countries and states
HCP	Healthcare professionals
HIC	High-income countries
HIV/AIDS	Human immunodeficiency virus/acquired immunodeficiency syndrome
HL	Health literacy
HSR	Health Systems Research
ICDM	Integrated Chronic Disease Management
ICU	Intensive care units
IHR	International Health Regulations
IPC	Infection, prevention, and control
LGTBIQ	Lesbian, gay, transgender, bisexual, intersex, and queer
LMIC	Low - and middle-income countries
MERS	Middle East Respiratory Syndrome
MHL	Media Health Literacy
MSM	Men who have sex with men
NAAT	Nucleic acid amplification technology
NTD	Neglected tropical diseases
OECD	The Organization for Economic Co-operation and Development
PAVM	Partnership for African Vaccine Manufacturer
PCSD	Policy Coherence for Sustainable Development
PLWH	People living with HIV
PPE	Personal protective equipment
PrEP	Pre-exposure prophylaxis
RCDC	Royal Centre for Disease Control
SARS	Severe acute respiratory syndrome

SDG	Sustainable Development Goals
TB	Tuberculosis
UNAIDS	The Joint United Nations Programme on HIV and AIDS
UHC	Universal health coverage

1 Sustainable Health in Low and Middle Income Countries; Focus on SDG Target 3.3, 3b, 3c and 3d

Abstract

A generation of children is at risk after missing out on school, essential vaccinations as well as food crisis and loss of parents from COVID-19. The impact of COVID-19 in Low-and-middle income countries is immense and the estimated losses will take years to recover and or permanently affect some of the struggling economies. According to the WHO World Malaria Report 2021, moderate disruptions were observed in malaria endemic countries, despite a 17% increase in malaria cases globally and two thirds of mortality cases in 2020 attributing to COVID-19 related disruptions. According to the 2021 UNAIDS report, eight countries were able to reach the 90-90-90 target while 19 countries managed to reach at least one of the targets. COVID-19 negatively affected identification of new TB cases, treatment and follow up of existing cases, as well as control and prevention of spread of TB within affected communities. The reporting of TB cases by countries was also affected with shortfalls in reporting observed in different regions of the world including India (41%), Philippines (12%), China (8%), Indonesia (14%). This chapter systematically analyses three communicable disease priorities, Malaria, TB and HIV, its targets and indicators and how these have been affected by the COVID-19 pandemic. Despite strides being made in some areas SDG3.3, the COVID-19 pandemic has caused interruptions that have and will considerably affect vaccination coverage as well as the progress that was made. The overarching question remains, can LMIC be able to build sustainable health systems?

Keywords

Sustainable health · COVID-19 · Inequalities · Sustainable development goals (SDG) · SDG3 · Low and Middle Income Countries (LMIC) · Vulnerable populations · Pandemics · HIV/AIDS · TB

1.1 Introduction

COVID-19 has highlighted the vulnerabilities in health systems and lack of coordination and international cooperation including the immense inequalities in response and access to treatment.

A generation of children is at risk after missing out on school, essential vaccinations as well as food crisis and loss of parents from COVID-19. The impacts of COVID-19 in LMIC are immense and the estimated losses will take years to recover and or permanently affect some of the struggling economies.

Inequalities between rich and poor economies have never been so highly visible. The response

F. A. Mphande, *Sustainable Health in Low and Middle Income Countries*, Sustainable Development Goals Series, https://doi.org/10.1007/978-981-99-4254-1_1

to the pandemic has shown exclusion of LMIC in many ways, including poor vaccine access, access to information and access to required therapies. The onset of omicron resulted in immediate closure of borders for most African countries, affecting travel and immensely affecting imports and exports from these countries. Despite the virus already circulating in several high income countries, the response highlighted double standards in pandemic response between LMIC and high income countries.

Considering this, how can LMIC sustain the health and wellbeing of their populations? Is there a possibility that these countries will attain SDG by 2030?

The last 2 years have shown the world how countries' health systems struggle to survive and how economic prowess determines the ability for countries to attain definitive goals when it comes to health decision making, public health measures, disease prevention and control. The pandemic started out as an infectious disease problem but in the end profound effects on social and economic aspects, health systems and communities were observed. Disparities between the rich and the poor grew and became more obvious as prevention and control measures were rolled out. While high income countries struggled to manage an infectious disease of such magnitude, low and middle income countries who have greater experience is infectious disease outbreaks were able to grasp the challenges that lay ahead regarding controlling and preventing the disease in their respective communities and setting. With abundant resources, high income countries were able to quickly produce vaccines that were widely distributed among the high income countries, and the "crumbs" were shared with low income countries (Wouters et al. 2021). To date vaccine coverage between high and low income countries is skewed with high income countries having attained compared to low income countries. According to WHO as of May, 2022, only 58 countries globally, almost all high income countries, had vaccinated 70% of their population, while approximately 1 billion in low income countries are still unvaccinated (https://www.who.int/emergencies/diseases/novel-coronavirus-2019/covid-19-vaccines; https://www.who.int/campaigns/vaccine-equity). Only 37% of health workers on low income countries were vaccinated, increasing the risk of acquiring infection while on duty (https://www.who.int/campaigns/vaccine-equity). With some of the vaccines shared through COVAX, some with less than a year to expiration date, populations in LMIC became sceptical and misinformation started to spread (Table 1.1). While some populations believe that vaccines are important, the spread of misinformation led to vaccine hesitancy, which paved way to many refusing to get the COVID-19 vaccination (Arce et al. 2021; Troiano and Nardi 2020; Wouters et al. 2021). Some of the factors that contributed to vaccine hesitancy include; ethnicity, working status, religion, politics, gender, age, education, income (Anthony 2012; Blair et al. 2017; Jegede 2007; Martinez-Bravo and Stegmann 2022; Troiano and Nardi 2020).

Despite the hesitancy towards the COVID-19 vaccine in certain populations, other vaccination campaigns were making progress before the start of the pandemic (Global Polio Eradication Initiative 2020; Ho et al. 2022; Sbarra et al. 2021; Shet et al. 2022; UNICEF 2021; World Health Organisation(WHO) Geneva n.d.; World Health Organization 2020b). The onset of the COVID19 pandemic disrupted not only economies and social wellbeing, but also public health interventions including, vaccination programs for various vaccine preventable diseases, as well as on going health campaigns for epidemic diseases including HIV/AIDS, malaria, tuberculosis (TB), neglected tropical diseases (NTD), hepatitis, waterborne diseases, and other communicable diseases (Global Polio Eradication Initiative 2020). Low- and middle-income countries that carry most of the burden of these infectious diseases struggled to sustain the programs during the pandemic, possibly due to lack of resources to support the programs as well as COVID-19 prevention and control measures affecting implementation of the programs (World Health Organisation(WHO) Geneva 2021b).

Table 1.1 2022 World Bank classification of countries according to their economies (GNI per capita). (Source: https://datahelpdesk.worldbank.org/knowledgebase/articles/906519-world-bank-country-and-lending-groups)

Lower-income economies ($1085 OR less)	Lower-middle income economies ($1086 to $4255)	
Afghanistan	Angola	Mauritania
Burkina Faso	Algeria	Micronesia, Fed. States
Burundi	Bangladesh	Mongolia
Central African Republic	Benin	Morocco
Chad	Bhutan	Myanmar
Congo, Dem. Rep	Bolivia	Nepal
Eritrea	Cabo Verde	Nicaragua
Ethiopia	Cambodia	Nigeria
The Gambia	Cameroon	Pakistan
Guinea	Comoros	Papua New Guinea
Guinea-Bissau	Congo, Republic	Philippines
Korea, Democratic People's Republic	Côte d'Ivoire	Samoa
Liberia	Djibouti	São Tomé and Principe
Madagascar	Egypt, Arab Rep.	Senegal
Malawi	El Salvador	Solomon Islands
Mali	Eswatini	Sri Lanka
Mozambique	Ghana	Tanzania
Niger	Haiti	Tajikistan
Rwanda	Honduras	Timor-Leste
Sierra Leone	India	Tunisia
Somalia	Indonesia	Ukraine
South Sudan	Iran, Islamic Rep	Uzbekistan
Sudan	Kenya	Vanuatu
Syrian Arab Republic	Kiribati	Vietnam
Togo	Kyrgyz Republic	West Bank and Gaza
Uganda	Lao PDR	Zimbabwe
Yemen, Rep.	Lebanon	
Zambia	Lesotho	

The overarching question remains, can LMIC be able to build sustainable health systems?

The world population as of 2022, stands at 7.84 billion https://data.worldbank.org/indicator/SP.POP.TOTL, 3.36 billion people live in Lower middle income countries https://data.worldbank.org/indicator/SP.POP.TOTL?locations=XN, while 701,926,973 live in low income countries https://data.worldbank.org/indicator/SP.POP.TOTL?locations=XM.

2.5 billion in upper middle income countries https://data.worldbank.org/indicator/SP.POP.TOTL?locations=XT and 1.24 billion live in high income countries https://data.worldbank.org/indicator/SP.POP.TOTL?locations=XD . Upper and middle income countries make up 3.74 billion of the world population, while lower income and lower-middle income economies make up 4.06 billion (51.81%) of the global population.

According to the WHO World Malaria Report 2021, moderate disruptions were observed in malaria endemic countries, despite a 17% increase in malaria cases globally and two thirds of mortality cases in 2020 attributing to COVID-19 related disruptions. The disruptions affected areas including ITN campaigns and distribution (World Health Organisation(WHO) Geneva 2021b).

As populations worldwide face the COVID-19 pandemic, while fighting existing pandemics and other emerging and re-emerging infectious diseases as well as increase in non-communicable diseases, SDG3, "Ensure healthy lives and well-being for all at all ages" comes back into focus (https://SDG.un.org/goals/goal3). Several targets have been set to meet SDG3, this book will focus on the following targets;

Target 3.3: By 2030, end the epidemics of AIDS, tuberculosis, malaria and neglected tropical diseases and combat hepatitis, water-borne diseases, and other communicable diseases (https://SDG.un.org/goals/goal3) (Fig. 1.1).

- Target 3.B Support the research and development of vaccines and medicines for the communicable and noncommunicable diseases that primarily affect developing countries, provide access to affordable essential medicines and vaccines, in

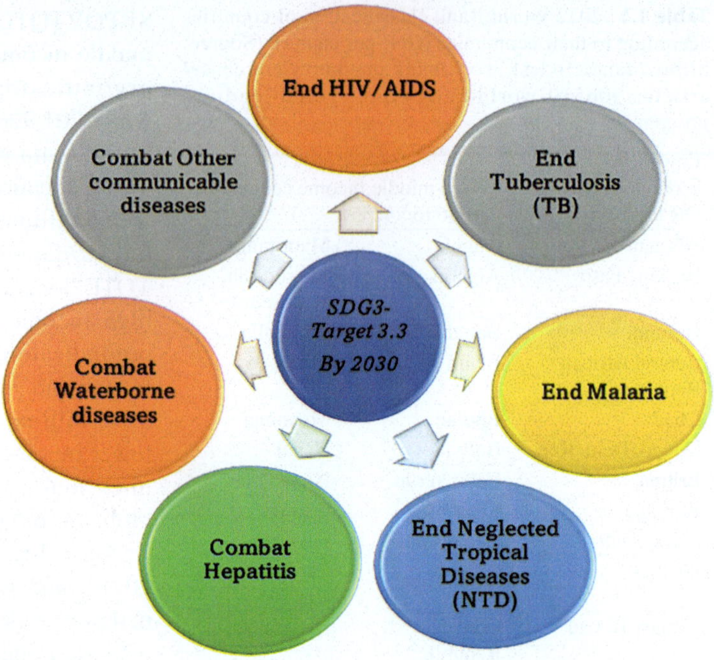

Fig. 1.1 SDG3—Target 3.3 Ending pandemics of AIDS, TB, malaria and NTDS and combating communicable diseases including waterborne diseases and hepatitis by the year 2030

accordance with the Doha Declaration on the TRIPS Agreement and Public Health, which affirms the right of developing countries to use to the full the provisions in the Agreement on Trade Related Aspects of Intellectual Property Rights regarding flexibilities to protect public health, and, in particular, provide access to medicines for all (https://www.wto.org/english/thewto_e/minist_e/min01_e/mindecl_trips_e.htm) (https://SDG.un.org/goals/goal3).

- Target 3.C Substantially increase health financing and the recruitment, development, training, and retention of the health workforce in developing countries, especially in least developed countries and small island developing States (https://SDG.un.org/goals/goal3).
- Target 3.D Strengthen the capacity of all countries, in particular developing countries, for early warning, risk reduction and management of national and global health risks (https://SDG.un.org/goals/goal3)(Fig. 1.2).

1.1.1 Post COVID-19 Disruptions: Globally

The Covid-19 pandemic caused disruptions globally both socially and economically as well as in healthcare and public health campaigns globally. These disruptions affected existing programs in various ways, causing losses in gains that were achieved previously towards meeting SDG. Some of the programs affected include, HIV/AIDS campaigns, malaria, TB, neglected tropical diseases (NTDs) as well as many vaccine-preventable disease programs.

1.1.1.1 END AIDS by 2030

END AIDS campaign was set up to combat HIV and various strategies were put in place through different institutions globally. These strategies require different approaches that can help to reduce the number of cases and deaths from HIV and AIDS. These approaches could involve either treatment and or epidemiological and global health. The approaches have to include both the grassroots at regional level national and global level, critical attention should be paid to social and cultural factors that may affect the response to treatment of the disease (Eisinger and Fauci

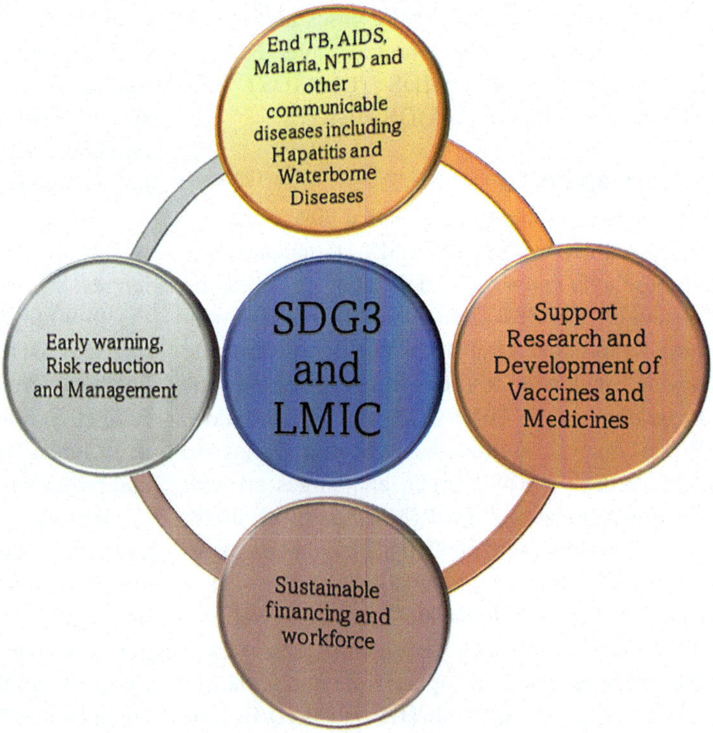

Fig. 1.2 Research and development needs in LMIC as part of SDG3

2018). To combat HIV some of the strategies that have been conducted include implementation of biomedical research advances aimed at treatment and prevention. Despite the availability of these approaches not every country has been able to achieve the required outcomes. Many countries with limited resources have found themselves wanting but failing to achieve the research that is required to develop vaccines and treatments that can help serve their populations. As such these countries depend on high income countries to provide for the needed treatment and prevention measures such as medicines as well as preventive therapy. In the past decade over 30 antiretroviral drugs have been approved for use for treatment of HIV and AIDS globally. Despite this, there are still implementation challenges that have affected adherence and treatment. Some of these challenges include drug toxicity, inconsistent uptake of treatment, drug resistance, pill fatigue limited access to treatment as well as the decision as to when to start treatment. This is an enormous challenge for LMIC to reduce the burden of HIV and AIDS. To achieve the End HIV and AIDS 2030 campaign, LMIC need to ensure that they are able to sustain treatment and prevention measures required to reduce the number of cases and deaths from HIV and AIDS (Cohen et al. 2016; Fauci and Marston 2015; Rodger et al. 2016). Antiretroviral treatment has allowed for reduced transmission of HIV from an infected partner to an uninfected partner and as well as reduction of transmission between mother and child. Use of antiretroviral treatment as well as preventive therapy such as PrEP, are underutilised in certain parts of the world as such there is need for access to such treatment for vulnerable populations in limited resource countries around the world (Cohen et al. 2016; Marcus et al. 2017).

The goal of this target is to end HIV by the year 2030 (UNAIDS 2016). To achieve this, one of the strategies that was employed was the Fast Track Strategy. In this strategy several targets were set up, the first part were goals targeting to be reached by 2020 and another set of targets to be reached by the year 2030. The fast track strategy plan targeted that by the year 2020, the focus will be on treatment, through the 90-90-90 scheme; reduction of new HIV cases and put an end to discrimination (Joint United Nations Programme ON HIV/AIDS (UNAIDS) n.d.; Marinda et al. 2020; UNAIDS 2016).

1.1.1.2 90-90-90

The 90-90-90 schemes goal is that 90% of people living with HIV know their HIV status, and 90% of those who know their HIV status are accessing treatment, and 90% of people attaining treatment have suppressed viral load) this is to eliminate new infections in children (Joint United Nations Programme ON HIV/AIDS (UNAIDS) n.d.). While many countries have been working to achieve the 90-90-90 target, challenges are still there to achieve this goal (Marinda et al. 2020; Ntata 2007). Several challenges have been experienced both on the patient side as well as countries, including diagnostic gaps as well as difficulties to estimate diagnostic rate (Choi 2020; Ellman 2015; Khopkar-Kale and Kulkarni 2021). People living with HIV (PLWH) in limited resource setting struggle with access to the required treatment, not because it is not available but due to proximity of treatment centres where they can access treatment as such, PLWH require transportation to access the facilities which provide ART (Ellman 2015). Transportation costs as well as the transportation itself to get to the place where they can access the treatment becomes a challenge. On the other hand, time required to access the treatment facilities could result in reluctance to travel to treatment facilities especially for those who are healthy but living with HIV. Even though some people will be able to access treatment centres, stigmatising attitudes from health personnel at the treatment facilities may deter PLWH from getting treatment (Ellman 2015; Jones et al. 2020). Apart from patients, countries are also struggling to provide ART to their patients. This can be attributed to several factors including loss of patients to follow up, patients reporting very late to the hospital when the disease is more advanced, and lack of sustainable funding (Ellman 2015). Some of the ART programs are supported through donor funding; lack of such funding could lead to lack of supply of ART. With universal treatment coming up as a conversation, most health systems in developing countries will struggle to meet this course in part due to some of the issues highlighted. The pressure to treat every patient regardless of their CD4 count as well as whether they are symptomatic or not we will increase the burden on health systems and could overwhelm health systems most of which are already struggling. Despite the challenges, 8 countries affected by HIV have attained the 90-90-90 by 2020 target (UN Joint Programme on HIV/AIDS (UNAIDS) 2021). One area that has been struggling is to estimate the number of PLWH. While estimation of PLWH is still work in progress, estimates of people receiving ART and reduction of new HIV cases are on track.

According to the 2021 UNAIDS report, eight countries were able to reach the 90-90-90 target and 19 countries managed to reach at least one of the targets. The countries that were able to reach the 90-90-90 targets include Eswatini, Switzerland, Rwanda, Qatar, Botswana, Slovenia, Uganda, and Malawi. Eswatini was ranked highest on reaching the target (with 97% viral suppression level). The success was attributed to strong political leadership, availability of resources where the government was able to allocate fans towards the response to HIV/AIDS, as well as understanding of cultural behaviour. The understanding of cultural behaviour helped in soliciting response which was culturally acceptable, and this helped the society to embrace the pandemic and the response efforts. Community mobilisation was another strategy that was employed, and this worked to their advantage in reaching 90-90-90 target. Although not many countries were able to attain the 90-90-90 target by the year 2020, 84% (31.6 million) of people living with HIV knew their HIV status, 73% (27.4 million) were accessing treatment and 66% (24.8 million) attained viral suppression (UN Joint Programme on HIV/AIDS (UNAIDS) 2021). Conversely the 2022 UNAIDS report has shown faltering progress including shrinking resources as well as widening inequality gaps within and among countries. The report highlighted insufficient investment and millions of HIV related deaths and new infections threatening the End AIDS by 2030 campaign. In 2021,

1.5 million new HIV cases and 650,000 AIDS related deaths were recorded despite the availability of tools to prevent, detect and treat opportunistic infections. Tech monopoly, austerity, clamping down on marginalized populations and non-inclusiveness were highlighted as some of the challenges towards attaining the 2030 goal (Joint United Nations Programme on HIV/AIDS 2022; UNAIDS n.d.).

Despite the achievements attained, the COVID-19 pandemic affected most of the responses in LMIC as well as globally. The responses to COVID-19 highlighted inequalities that existed between high income countries and LMIC. These inequalities have had detrimental effect in the response to HIV pandemic and other public health issues including COVID-19. Although LMIC work together with higher income countries to attain better response and treatment to HIV, the inequalities that exist between these two social economic groups have shown that HIV response is still a challenge in many LMIC. Gaps in HIV response have been observed and resource limited countries, especially in children who were unable to go to school and or dropped out from the schools due to at the COVID-19 pandemic (Joint United Nations Programme on HIV/AIDS 2022). These children especially girls were at high risk of contracting HIV. Inequalities globally resulted in gaps in testing and treatment of HIV as well as insufficient action in the fight against HIV. Inequalities among highly marginalised populations, those with no access to healthcare and disparities according to age and sex, were observed. This led to poor response in many countries and this in turn affected the outcomes of the end HIV campaign. Poverty and education level were also highlighted as factors exacerbating inequalities within LMIC. Disparities between the rich and the poor as well ad rural and urban populations is another gap affecting populations globally. It was observed that within the same country, while the wealthy could easily access better healthcare, better testing facilities and treatment; the poor struggled immensely (UN Joint Programme on HIV/AIDS (UNAIDS) 2021). Education and wealth played a substantial role in how people respond to family planning, as well as treatment of HIV. People living in urban areas had higher chances of accessing treatment compared to those in rural areas highlighting the issues of access. Apart from the pandemic other social economic challenges within different in LMIC have encouraged migration of people from low income countries to high income countries. Migration may contribute to disruption of HIV treatment due to limited access to healthcare due to the migrants migration status. Normalisation of plights and challenges in different populations has allowed for growth in treatment and response disparities. While it may seem normal for certain populations to be poor, struggle with disease and have little or no access to descent and needed treatment, it is not normal and it shouldn't be so. The mentality of the haves and the have nots, the deserving and the undeserving, have exacerbated inequality in both access and response to public health challenges and these inequalities have driven response and treatment to HIV in LMIC to their lowest. It is therefore important to address these existing inequalities including healthcare access, gender and racial inequalities as well as denial of people's human rights as a major obstruction to achieving HIV response globally.

1.1.1.3 End TB

Through the End TB by 2030 campaign, efforts are being made globally to fight the pandemic. The End TB by 2030 campaign builds on four principles, these are; government stewardship and accountability with monitoring and evaluation, strong coalition with civil society organizations and communities, protection and promotion of human rights, ethics and equity, and adaptation of the strategy and targets at country level with global collaboration (World Health Organization 2015). For LMIC, TB and COVID-19 pandemic were a huge blow to the already struggling health systems in most of these countries (Alagna et al. 2020). The similarities between COVID-19 and TB are staggering, with both diseases having a great potential of overwhelming the health system, requiring rapid diagnosis, having limited knowledge of individuals susceptibility to infections, poor coordination both at national and international level and requiring public awareness for disease prevention and control (Alagna et al. 2020; Wang et al. 2020). Acquiring the vital supplies to manage the pandemic including PPE, medical ventilators as well as having enough equipment and space to take care of sick patients in the ICU were a great challenge (Alagna et al. 2020; Wang et al. 2020). The COVID-19 pandemic has affected the campaign in many different ways, 1 year after the COVID-19 pandemic started, it was observed that there was a decline in the diagnosis and treatment of TB infections (Adewole 2020; Battista Migliori et al. 2020; Klinton et al. 2021; Migliori et al. 2021; Pillay et al. 2021; Souza et al. 2022; Wu et al. 2020). The decline affected the reported numbers of people diagnosed with TB as well as the access to treatment for people affected with TB (Buonsenso et al. 2021; Walker et al. 2018; Wang et al. 2020). These changes were attributed to COVID-19 pandemic as people were afraid to go to hospitals either because they did not want to get infected by COVID-19 or they didn't really want to know if they were infected with COVID-19 to avoid social stigma associated with the disease (Buonsenso et al. 2021). This negatively affected identification of new TB cases, treatment and follow up of existing cases, as well as control and prevention of spread of TB within the community. The reporting of TB cases by countries was affected with shortfalls in reporting observed in different regions of the world including countries such as India (41%), Philippines (12%), China (8%), Indonesia (14%). TB reporting in Africa zone reduced by 2.5% considerably lower than other WHO regions (Klinton et al. 2021; World Health Organization 2020a). There were also deficits observed in case finding as well as treatment of TB. Resources including human resources, financial resources, and infrastructure available were diverted to deal with COVID-19. COVID-19 affected reporting of TB cases both at National and globally. The overall number of TB patients reported in 2020 was lower than expected high TB burden countries. Apart from reporting, TB care in LMIC was also affected, this was attributed to challenges to access care due to financial constraints and fear of contracting COVID-19 at healthcare facilities. After observing the reduction in TB diagnosis during the pandemic, an idea of integrated TB outreach into COVID-19 programmes was introduced and this was to try and combat both TB and COVID-19 so that the targets for TB are still met.

Trajman et al. reported that some of the challenges affecting TB programme during the COVID-19 pandemic were loss of healthcare workers to COVID-19 thereby affecting TB services (Trajman et al. 2022). The fear of contracting COVID-19 led to reduced access to TB services, as such those diagnosed with TB were not able to go to health services to access treatment due to fear of catching COVID-19. COVID-19 mortality was reported to increase among people who were affected with TB. TB notification was negatively affected by COVID-19 with notifications of new cases, deaths, and treatment, reduced in many countries as the resources were concentrated on COVID-19. Decrease in testing treatment and prevention coverage contributed negatively to the fight against TB in most of the affected countries (Trajman et al. 2022; World Health Organisation(WHO) Geneva 2021a). Apart from TB, the other challenge affecting LMIC are emerging and re-emerging infectious diseases most of which are of zoonotic origin.

1.1.1.4 The One Health Approach: Addressing Health threats across disciplines

The One Health approach (Sinclair 2019) supports global health security by improving coordination, collaboration, and communication. (Sinclair 2019). At the human-animal-environment interface there is need to address shared health threats such as zoonotic diseases, antimicrobial resistance, food safety and others. Over the past decade, country after country has implemented the One Health approach and demonstrated recognised benefits. However, to build sustainable One Health in these efforts, One Health champions and implementers need to collect and provide government decision-makers with country-level data on One Health's impact to help justify policy decisions and resource allocations. Due to the broad, often seemingly all encompassing, nature of One Health in promoting synergies of multiple disciplines and sectors, the One Health community has faced difficulties in determining specific One Health impact indicators for formally evaluating One Health successes. The benefits of One Health approach are many, but at the helm of it is understanding the interaction between animal as reservoirs of disease and humans as susceptible hosts. Implementation of One health varies from country to country depending on the setting, animal hosts as well as pathogen spillover events. With zoonotic diseases contributing to 70% of the emerging and re-emerging diseases, One health is an important aspect to consider if the world is to achieve sustainable health.

1.2 Vulnerable Populations

1.2.1 Research and Development of Vaccines in LMIC

Research and development (R and D) in LMIC is an area that has struggled and is still struggling to pick up. Despite availability of well trained and qualified researchers from LMIC, lack of up-to-date facilities, infrastructure, and funding have been some of the major setbacks in developing sustainable trade research. A substantial proportion of individuals trained in high income count who want to remain in research tend to remain in the countries where they were trained or move to other high-income countries where infrastructure is available. This have negatively contributed to R and D in most LMIC. While some countries are able to conduct research and have reputable research centers, most of these centers are still affiliated to funders from high income countries affecting the ability of host countries to have total control on their needs. In such cases the research is based on the policy of the funders which most often rarely focus on the research priorities of the affected countries but promotes the needs of the funding organizations.

In terms of infectious diseases, LMIC have been used as a source of samples for research most often with little or no recognition of local researchers. This has left many in LMIC as a link to the host country, but unable to be part of the outcomes including acknowledgment as part of list of authors in publications and or research talks (Chaccour 2018; Kelaher et al. 2016; Pingray et al. 2020).

High conference fees and visa requirements have also shut out many LMIC researchers from the global arena. It is at such conferences that resolutions are passed for diseases affecting LMIC, where consortia and collaborations are birthed. It remains shocking hat such exclusion criteria by high income countries are entertained in the twenty-first century and yet the world is still talking about SDG. How can SDG be achieved in the midst of such inequitable distribution of knowledge? It is high time that these policies are revised to suit the twenty-first century otherwise the gap of inequality is too wide to be talking about SDG. For Sustainable health in LMIC there is need for revision of systemic exclusion of these countries in decisions that affect the health of their populations.

The groundbreaking announcement of South Africa for sequencing of omicron resulted in blocking several African countries from entry into Europe, America, and Asia yet the discovery of other variants in high income countries did not get such a response (Mehta 2022; Schermerhorn et al. 2022). Undermining research outcomes from LMIC is testament of such systemic behavior that discoveries can only be made in the global north. Unless this attitude is addressed, attaining SDG is still a long way.

The establishment of centers of research excellence in LMIC focusing on promoting research and technology would be essential (Nkengasong et al. 2017). These centers will not only focus on what is required in these countries but combat the existing public health challenges that are neglected but affect Billions in LMIC. The establishment of the mRNA vaccine technology transfer Hub in South Africa, is a "historic initiative that comes at a critical time" in the global fight against the virus (WHO 2022).

Creating Centers of research excellence in LMIC would assist In driving some of the research needs at country, regional and global level.

1.2.2 The Effect of the COVID-19 Pandemic on Global Public Health Campaigns

Despite strides being made in some areas for target 3.3, the COVID-19 pandemic has caused interruptions that have and will considerably affect vaccination coverage as well as the progress that was made, for example in reducing malaria cases (Pillay et al. 2021). Vulnerable populations who were already struggling to access their healthcare needs before the pandemic, could face even greater challenges at present and in the years to come post-pandemic. This book will focus on what is the progress on attaining the targets? How has the COVID-19 pandemic affected the progress? What can be done to strengthen capacities in all countries? What can or should be done to ensure sustainable health systems in developing countries including early warning systems, risk reduction and management of global and national health risks?

Some of the impacts of COVID-19 as reported by the United Nations Department of Social Affairs, include an additional 101 million children have fallen below the minimum reading proficiency level, potentially wiping out two decades of education gains. The global extreme poverty rate rose for the first time in over 20 years, and 119–124 million people were pushed back into extreme poverty in 2020. Women have faced increased domestic violence, child marriage is projected to rise after a decline in recent years, and unpaid and underpaid care work is increasingly and disproportionately falling on the shoulders of women and girls, impacting educational and income opportunities and health. The pandemic has also brought immense financial challenges, especially for developing countries—with a significant rise in debt distress and dramatic decreases in foreign direct investment and trade (United Nations 2021).

1.3 Conclusion

The overarching question remains, can LMIC be able to build sustainable health systems?

The impact of COVID-19 on the various public health campaigns including vaccination programs have shown how health systems in LMIC and well as in well income countries can be overwhelmed by an infectious disease. While trying to recover from the impacts of the pandemic, it is worth considering the lessons learned and how countries can adapt themselves to manage infectious diseases with the available resources.

References

Adewole OO (2020) Impact of COVID-19 on TB care: experiences of a treatment centre in Nigeria. Int J Tuberc Lung Dis 24(9):981–982. https://doi.org/10.5588/IJTLD.20.0418

Alagna R, Bessozi G, Codesca LR, Gori A, Migliori GB, Raviglione M, Cirillo DM (2020) Celebrating World Tuberculosis Day at the time of COVID-19. Eur Respir J 55:2000650. https://doi.org/10.1183/13993003.00650-2020

Anthony R (2012) The CIA' s vaccination ruse. J Public Health Policy 33(August):387–389. https://doi.org/10.1057/jphp.2012.37

Arce JSS, Warren SS, Meriggi NF, Scacco A, Mcmurry N, Voors M et al (2021) COVID-19 vaccine acceptance and hesitancy in low- and middle-income countries. Nat Med 27:1385–1394. https://doi.org/10.1038/s41591-021-01454-y

Battista Migliori G, Min Thong P, Akkerman O, Alffenaar J-W, Álvarez-Navascués F, Mohamed Assao-Neino M et al (2020) Emerging infectious diseases WWW worldwide effects of coronavirus disease pandemic on tuberculosis services, January-April 2020 the Philippines (M. Calnan). Barts Health National Health Service Trust 26(11):2709–2712. https://doi.org/10.3201/eid2611.203163

Blair RA, Morse BS, Tsai LL (2017) Social science & medicine public health and public trust: survey evidence from the Ebola virus disease epidemic in Liberia. Soc Sci Med 172:89–97. https://doi.org/10.1016/j.socscimed.2016.11.016

Buonsenso D, Iodice F, Sorba Biala J, Goletti D (2021) COVID-19 effects on tuberculosis care in Sierra Leone. Pulmonology 27(1):67. https://doi.org/10.1016/J.PULMOE.2020.05.013

Chaccour J (2018) Authorship trends in The Lancet Global Health: only the tip of the iceberg? Lancet Glob Health 6(5):e497. https://doi.org/10.1016/S2214-109X(18)30110-4

Choi JY (2020) The HIV Care Cascade in Korea: status of UNAIDS 90-90-90 targets. https://doi.org/10.3346/jkms.2020.35.e55

Cohen MS, Chen YQ, McCauley M, Gamble T, Hosseinipour MC, Kumarasamy N et al (2016) Antiretroviral therapy for the prevention of HIV-1 transmission. N Engl J Med 375(9):830. https://doi.org/10.1056/NEJMOA1600693

Eisinger RW, Fauci AS (2018) Ending the HIV/AIDS Pandemic. Emerg Infect Dis 24(3):413. https://doi.org/10.3201/EID2403.171797

Ellman T (2015) Demedicalizing AIDS prevention and treatment in Africa. N Engl J Med 372(4):303–305. https://doi.org/10.1056/NEJMp1414213

Fauci AS, Marston HD (2015) Ending the HIV–AIDS pandemic — follow the science. N Engl J Med 373(23):2197–2199. https://doi.org/10.1056/NEJMP1502020/SUPPL_FILE/NEJMP1502020_DISCLOSURES.PDF

Global Polio Eradication Initiative (2020) UNICEF and WHO call for emergency action to avert major measles and polio epidemics. World Health Organization. https://www.who.int/news/item/06-11-2020-unicef-and-who-call-for-emergency-action-to-avert-major-measles-and-polio-epidemics. Accessed 26 Mar 2023

Ho LL, Gurung S, Mirza I, Nicolas HD, Steulet C, Burman AL et al (2022) Impact of the SARS-CoV-2 pandemic on vaccine-preventable disease campaigns. Int J Infect Dis 119:201–209. https://doi.org/10.1016/j.ijid.2022.04.005

Jegede AS (2007) What Led to the Nigerian Boycott of the Polio Vaccination Campaign? 4(3):e73. https://doi.org/10.1371/journal.pmed.0040073

Joint United Nations Programme on HIV/AIDS. (2022) UNAIDS Global AIDS Update 2022. UNAIDS Global AIDS Update 2022, 2022, 1–22

Joint United Nations Programme ON HIV/AIDS (UNAIDS). (n.d.). Joint United Nations Programme ON HIV/AIDS (UNAIDS) 90-90-90 An ambitious treatment target to help end the AIDS epidemic [Online] UNAIDS, Geneva (2014). http://www.unaids.org/sites/default/files/media_asset/90-90-90_en_0.pdf

Jones HS, Floyd S, Stangl A, Bond V, Hoddinott G, Pliakas T et al (2020) Association between HIV stigma and antiretroviral therapy adherence among adults living with HIV: baseline findings from the HPTN 071 (PopART) trial in Zambia and South Africa. Trop Med Int Health 25(10):1246–1260. https://doi.org/10.1111/tmi.13473

Kelaher M, Ng L, Knight K, Rahadi A (2016) Equity in global health research in the new millennium: trends in first-authorship for randomized controlled trials among low- and middle-income country researchers 1990-2013. Int J Epidemiol 45(6):2174–2183. https://doi.org/10.1093/ije/dyw313

Khopkar-Kale P, Kulkarni S (2021) HIV-2: still off the radar for India's 90-90-90 targets. Indian J Med Res 154(1):24. https://doi.org/10.4103/IJMR.IJMR_828_19

Klinton JS, Heitkamp P, Rashid A, Faleye BO, Win Htat H, Hussain H et al (2021) One year of COVID-19 and its impact on private provider engagement for TB: A rapid assessment of intermediary NGOs in seven high TB burden countries. J Clin Tuberc Other Mycobact Dis 25:100277. https://doi.org/10.1016/j.jctube.2021.100277

Marcus JL, Hurley LB, Nguyen DP, Silverberg MJ, Volk JE (2017) Redefining Human Immunodeficiency Virus (HIV) preexposure prophylaxis failures. Clin Infect Dis 65(10):1768. https://doi.org/10.1093/CID/CIX593

Marinda E, Simbayi L, Zuma K, Zungu N, Moyo S, Kondlo L et al (2020) Towards achieving the 90-90-90 HIV targets: results from the south African 2017 national HIV survey. BMC Public Health 20:1375. https://doi.org/10.1186/s12889-020-09457-z

Martinez-Bravo M, Stegmann A (2022) In vaccines we trust? The effects of the CIA's vaccine ruse on immunization in Pakistan. J Eur Econ Assoc 20(1):150–186. https://doi.org/10.1093/jeea/jvab018

Mehta S (2022) Covid-19 update: omicron variant – a new emerging threat. Rocz Panstw Zakl Hig 73(1):13–16. https://doi.org/10.32394/rpzh.2022.0198

Migliori GB, Thong PM, Alffenaar JW, Denholm J, Tadolini M, Alyaquobi F et al (2021) Gauging the impact of the COVID-19 pandemic on tuberculosis services: a global study. Eur Respir J 58(5):2101786. https://doi.org/10.1183/13993003.01786-2021

Nkengasong J, Djoudalbaye B, Maiyegun O (2017) A new public health order for Africa's health security. Lancet Glob Health 5:e1064–e1065. https://doi.org/10.1016/S2214-109X(17)30363-7

Ntata, P. R. T. (2007). Equity in access to ARV drugs in Malawi. SAHARA J …, 4(1), 564–574. doi:https://doi.org/10.1080/17290376.2007.9724818

Pillay Y, Pienaar S, Barron P, Zondi T (2021) Impact of COVID-19 on routine primary healthcare services in South Africa. S Afr Med J 111(8):714–719. https://doi.org/10.7196/SAMJ.2021.V111I8.15786

Pingray V, Ortega V, Yaya S, Belizán JM (2020) Authorship in studies conducted in low-and-middle income countries and published by reproductive health: advancing equitable global health research collaborations. Reprod Health 17(1):18. https://doi.org/10.1186/s12978-020-0858-7

Rodger AJ, Cambiano V, Bruun T, Vernazza P, Collins S, Van Lunzen J et al (2016) Sexual activity without condoms and risk of HIV transmission in serodifferent couples when the HIV-positive partner is using suppressive antiretroviral therapy. JAMA 316(2):171–181. https://doi.org/10.1001/JAMA.2016.5148

Sbarra AN, Rolfe S, Nguyen JQ, Earl L, Galles NC, Marks A et al (2021) Mapping routine measles vaccination in low- and middle-income countries. Nature 589(7842):415–419. https://doi.org/10.1038/s41586-020-03043-4

Schermerhorn J, Case A, Graeden E, Kerr J, Moore M, Robinson-Marshall S et al (2022) Fifteen days in December: capture and analysis of Omicron-related travel restrictions. BMJ Glob Health 7(3):e008642. https://doi.org/10.1136/bmjgh-2022-008642

Shet A, Carr K, Danovaro-Holliday MC, Sodha SV, Prosperi C, Wunderlich J et al (2022) Impact of the SARS-CoV-2 pandemic on routine immunisation services: evidence of disruption and recovery from 170 countries and territories. Lancet Glob Health 10(2):e186–e194. https://doi.org/10.1016/S2214-109X(21)00512-X

Sinclair JR (2019) Importance of a One Health approach in advancing global health security and the Sustainable Development Goals. Revue Sci Tech (International Office of Epizootics) 38(1):145–154. https://doi.org/10.20506/RST.38.1.2949

Souza M d R, Paz WS d, Sales VB d S, Jesus GFH d, Tavares D d S, Lima SVMA et al (2022) Impact of the COVID-19 Pandemic on the Diagnosis of Tuberculosis in Brazil: Is the WHO End TB Strategy at Risk? Front Pharmacol 13(June):1–10. https://doi.org/10.3389/fphar.2022.891711

Trajman A, Felker I, Alves LC, Coutinho I, Osman M, Meehan SA et al (2022) The COVID-19 and TB syndemic: the way forward. Int J Tuber Lung Dis 26(8):710–719. https://doi.org/10.5588/ijtld.22.0006

Troiano G, Nardi A (2020) Vaccine hesitancy in the era of COVID-19. (January)

UN Joint Programme on HIV/AIDS (UNAIDS). (2021) Global AIDS updates confronting inequities. 386

UNAIDS. (2016) 2016 United Nations political declaration on ending AIDS sets world on the fast-track to end the epidemic by 2030. http://www.unaids.org/sites/default/files/20160608_PS_HLM_Political_Declaration_final.pdf

UNAIDS. (n.d.) UNAIDS global AIDS update 2022

UNICEF. (2021) Lessons learned and good practices: country-specific case studies on immunization activities during the COVID-19 pandemic. https://www.unicef.org/documents/lessons-learned-and-good-practices-country-specific-case-studies-immunization-activities

United Nations (2021) Goal 3|Department of Economic and Social Affairs [Online]. UN Online. https://SDG.un.org/goals/goal3. Accessed 17 Mar 2023

Walker TM, Merker M, Knoblauch AM, Helbling P, Schoch OD, van der Werf MJ et al (2018) A cluster of multidrug-resistant Mycobacterium tuberculosis among patients arriving in Europe from the Horn of Africa: a molecular epidemiological study. Lancet Infect Dis 18(4):431–440. https://doi.org/10.1016/S1473-3099(18)30004-5

Wang X, Zhang X, He J (2020) Challenges to the system of reserve medical supplies for public health emergencies: reflections on the outbreak of the severe acute respiratory syndrome coronavirus 2 (SARS-CoV-2) epidemic in China. Biosci Trends 14(1):3–8. https://doi.org/10.5582/bst.2020.01043

WHO (2022) The mRNA vaccine technology transfer hub. World Health Organisation. https://www.who.int/initiatives/the-mrna-vaccine-technology-transfer-hub. Accessed 17 Mar 2023

World Health Organisation(WHO) Geneva. (2021a) Global Tuberculosis Report 2021. In News. Geneva

World Health Organisation(WHO) Geneva. (2021b) World Malaria Report 2021. Geneva

World Health Organisation(WHO) Geneva. (n.d.) GIN June 2020. https://www.who.int/publications/m/item/gin-june-2020. Accessed 16 Mar 2023

World Health Organization (2015) The End Strategy TB. End TB Strategy 53(9):1689–1699

World Health Organization. (2020a) Global Tuberculosis Report 2020. https://apps.who.int/iris/bitstream/handle/10665/336069/9789240013131-eng.pdf

World Health Organization. (2020b) Pulse survey on continuity of essential health services during the COVID-19 pandemic: interim report, 27 August 2020. In Interim report. COVID-19. Essential Health Services. https://www.who.int/publications/i/item/WHO-2019-nCoV-EHS_continuity-survey-2020.1

Wouters OJ, Shadlen KC, Salcher-konrad M, Pollard AJ, Larson HJ, Teerawattananon Y, Jit M (2021) Health Policy Challenges in ensuring global access to COVID-19 vaccines: production, affordability, allocation, and deployment. Lancet 397(10278):1023–1034. https://doi.org/10.1016/S0140-6736(21)00306-8

Wu Z, Chen J, Xia Z, Pan Q, Yuan Z, Zhang W, Shen X (2020) Impact of the COVID-19 pandemic on the detection of TB in Shanghai, China. Int J Tuberc Lung Dis 24(10):1122–1124. https://doi.org/10.5588/IJTLD.20.0539

2 No-One Left Behind: A Holistic Approach

Abstract

Research and innovation is key to the human population. Discovery of new methods and creation of new technologies is crucial for human development. Lack of research and innovation has impacts on human development. The onset of COVID-19 revealed just how much LMIC are disadvantaged systematically, and the lack of technological capacity to design and acquire vaccines. The gap in vaccine access was so enormous that it left questions on equitable access and universal health coverage. There is no sustainable health if there are still communities that have no access to healthcare, if there are still neglected diseases, if there are still vaccine inequities, if there are health inequalities and systematic exclusion of the most vulnerable and affected populations in decision making for their health and wellbeing.

Keywords

Low and Middle Income Countries (LMIC) · Research and innovation · Equitable access · COVID-19 · Systematic exclusion · Political unrest · Brain drain · Integrated health systems · Neglected tropical diseases

2.1 No-One Left behind

The research and innovation process is one area that has brought about advancement not only in medical research but also in various disciplines globally. The benefits of research and innovation can be seen in major breakthroughs in public health including the production of modern vaccines, diagnostics, treatment, and therapies (Afsahi et al. 2018; Barrett and Croft 2012; Fang et al. 2022).

What Is Research and Innovation?
Research can be defined as a discipline that involves a methodological study and analysis of various factors, including biological, epidemiological, and genetic factors and the use of appropriate statistical tools to produce evidence based outcomes.

Innovation uses outcomes from research to produce new and improved technologies that could enhance research and or improve wellbeing (Nass et al. 2009).

Need for Research and Innovation from LMIC for LMIC
Research and innovation is key to the human population. Discovery of new methods and creation of new technologies is crucial for human development. Lack of research and innovation has impacts on human development. High income countries are known for their grand investments

F. A. Mphande, *Sustainable Health in Low and Middle Income Countries*, Sustainable Development Goals Series, https://doi.org/10.1007/978-981-99-4254-1_2

in research and development which in turn have resulted in the development of various technologies in the last centuries. The past two decades have produced enormous technological advancements that have changed the social, economic as well as the health environment globally (Fang et al. 2022; Quick et al. 2016; Schnee et al. 2016). Advancement in digital health, new technologies that inform health, equipment and therapies that have improved health and wellbeing of many have been reported and implemented. Despite these advancements not everyone is developing at the same pace. While high income countries (HIC) have further advanced technologically, LMIC are lagging behind at great proportions. Why is this the case?

Several factors have been attributed to these gaps, apart from economic challenges, limited access, and governance are some of the major factors not to be undermined.

2.1.1 Equitable Access

A simple example to illustrate access in LMIC could be acquisition of reagents and or equipment for scientific research. While in HIC this can be done by a simple phone call and a product can be delivered within 24 h, in LMIC, it will take 3 months or more to receive the same product with hefty import taxes and delivery charges. With this simple illustration not only is it difficult economically for most LMIC to acquire novel tools, but also the tools may not work to their optimal capabilities depending on various factors. Some of the factors include the manpower to operate the tools, the capacity to attain and maintain the tools, as well environmental factors that may affect operation of the tools (Hasan et al. 2021; Rosa et al. 2022). One option that could be of great importance will be the ability of LMIC to manufacture tools that are essential for their populations but also that can function optimally and can be easily accessed and maintained within their settings (Hotez and Ferris 2006). Most LMIC act as suppliers of research material including DNA, blood, and other samples that are collected and sent to other countries with the capability to analyze them further. If LMIC are able to carry out advanced research within their territories using materials and resources that are readily accessible within their territories, this may greatly improve research and innovation within the region (Addo-Atuah et al. 2020; Bangert et al. 2017).

2.1.1.1 COVID-19 and Access to Vaccines

While HIC had excess doses of COVID -19, LMIC were left at the mercy of donations from these countries. Low-and-middle income countries need vaccines for various vaccine preventable diseases that are endemic within the region. The idea of waiting for help from donors and high income countries to fund the needed vaccines has crippled the ability of LMIC to combat vaccine preventable diseases in the past five decades. The onset of COVID-19 revealed just how much LMIC are disadvantaged systematically, and the lack of technological capacity to design and acquire vaccines (Addo-Atuah et al. 2020). The gap in vaccine access was so enormous that it left questions on equitable access and universal health coverage (Eccleston-Turner and Upton 2021; Tagoe et al. 2021). The systematic exclusion that resulted from the lack of COVID-19 vaccine coverage left many in LMIC not to trust even the COVAX vaccine donations contributing to vaccine hesitancy in some parts (Abu-Odah et al. 2022; Deml and Githaiga 2022; Lee and You 2022; Tagoe et al. 2021). A study by Arce et al., showed that vaccine acceptance (primarily explained by an interest in personal protection against COVID-19) was higher in LMIC compared to other countries, with side effects as the most common reason for hesitancy (Arce et al. 2021). Despite these findings limited information is available on the actual access of COVID-19 vaccines in LMIC (Deml and Githaiga 2022).

Disparities in vaccine coverage have been observed not only geographically but also locally within a country. Vaccine coverage was observed to be higher in urban areas compared to rural areas in some LMIC including Cameroon, Ethiopia, India, Madagascar, Malawi, Myanmar,

Tanzania, Pakistan, and Vietnam while the Gambia, Mauritania, Nigeria, Eswatini, and Uzbekistan had higher vaccine coverage in rural areas (Ahmed Ali et al. 2022).

Apart from COVID-19, Monkeypox outbreak emerged in areas where it was not endemic (CDC 2022; Luna et al. 2022). This came in at the time that the world was recovering and making progress with COVID-19. The onset of Monkeypox has also revealed how response to an infectious disease is different depending on the affected populations. Correlation between income group and level of response was still obvious. The swift response to COVID-19 outbreak and Monkeypox outbreaks in high income countries is testament to this. While these infectious diseases can be managed and treated swiftly with advanced technology in the global North, the fact remains that if LMIC are left behind, these regions will remain reservoirs to such infections. If sustainable health is to be attained and indeed no-one is to be left behind this is where it must start, consolidated response to disease outbreaks and risks. If LMIC are left behind, no matter how advanced the technology sustainable health is but a dream.

Vaccine nationalism observed during the COVID-19 pandemic, cements the idea of prioritising the needs of ones' country over the others. Access to vaccines and other therapies could be attained by revisiting policies that are barriers to manufacturing due to intellectual property (IP) rights, these not only affect the manufacturing of medical products, but also expansion of this manufacturing in other parts of the world (Addo-Atuah et al. 2020; Guzel et al. 2021; Hotez and Ferris 2006). The current policies and frameworks surrounding vaccine and drug manufacturing are not sustainable for LMIC (Addo-Atuah et al. 2020; Bangert et al. 2017; Eccleston-Turner and Upton 2021; Peacocke et al. 2021).

2.1.1.2 Making Information Accessible for All

Access to technology requires access to reliable information. Despite the research material including human study subjects, blood, genetic material, and other forms of materials being obtained from LMIC, most of these countries have not benefited as much from the innovations and tools that have resulted from these studies. Access to research information includes access to current publications and ability to publish and present research findings. For example, while there are a number of well accomplished researchers in various disciplines in LMIC, most have to be affiliated to other organizations in high income countries to attain research grants, access state of the art equipment and join knowledge sharing networks to improve their knowledge (Addo-Atuah et al. 2020). With networks, researchers are able to access valuable information regarding advancements in their fields of interest as well as meet potential collaborators. Research outcomes and discoveries are usually presented in conferences, congresses, and many scientific gatherings around the world. Taking an example of tropical diseases, most of the published research is by researchers from non- tropical countries who carry out studies in tropical countries (Addo-Atuah et al. 2020; Chaccour 2018; Kelaher et al. 2016). While the work is published and presented, the countries where the research was carried out have minimal representation not only in the publication list but also access to the published material. Access to new ideas is also a challenge to many LMIC as most require visas to go to different countries. Visas have become such a big barrier for many researchers in LMIC to attend workshops, courses and conferences that could assist in attaining and knowledge exchange (Pai n.d.; Velin et al. 2021). https://www.internationalhealthpolicies.org/featured-article/time-to-rethink-global-collaboration-travel-restrictions-for-global-health-professionals/). This barrier has managed to widen the gap of knowledge between the global North and South leaving the global South with minimal chance of expanding public health research.

2.1.1.3 Systematic Exclusion*:*

Apart from poor access to information, systematic exclusion of the global South cannot be more obvious in research. For example, the research process involves several stages, from grant application, obtaining funding, publication of outcomes, to dissemination of the outcomes (congresses and conferences). LMIC have found themselves left out of grant applications for example by statements such as "one of the collaborating laboratories should be in global North", or the principal investigator should be from the global North", and many more similar statements that systematically exclude a certain group of researchers. Despite the availability of grants for LMIC, the number of grants is limited, as such the chance of attaining these grants are very slim. If a researcher obtains data and is keen to publish, publication fees are another area of exclusion. Most of the researchers from LMIC are deterred by the publication fees that are far more expensive and next to impossible for most researchers. Most of the high impact journals have publication fees between US $1000 and US $8000 or more which are next to impossible for most local researchers in LMIC to manage (Ghani et al. 2020; Newton 2020). This puts most of the researchers in LMIC at a disadvantage, where they are not able to fund their research, and if they manage to get data for publication they are not able to publish hence obtain grants. Despite having the data, the researchers may not be able to present it to audiences globally due to lack of access to relevant platforms (Addo-Atuah et al. 2020; Eccleston-Turner and Upton 2021).

Generally, LMIC have worked with infectious diseases for many years and have found ways to prevent and control the spread of these diseases within their settings. Most of these countries have endemic diseases that often occur periodically for example annually and or seasonal (Bangert et al. 2017; Leclerc et al. 2002; Seimenis 2012; World Health Organisation (WHO) 2016). No matter their frequency, most LMIC have often contained and managed the outbreaks and contained the spread preventing dissemination to other parts of the country and or region. Since these are localised outbreaks much information does not go beyond the victims and the health practitioners involved. This is not to undermine international support that has been offered for many decades but to discourage systematic exclusion. Lack of publications on how these outbreaks are managed in LMIC setting by local researchers means that countries within the region may not be able to access such helpful information that can help reduce the burden. No-one -one left behind is an initiative that can be adopted by these countries to support one another to prevent similar outbreaks within the region or build better surveillance networks that can help reduce the disease burden within the region (Bedriñana et al. 2021; Nkengasong et al. 2017).

2.1.1.4 Political Upheavals

Exclusion for some LMIC can be attributed to governance and other political issues that make it difficult to conduct research and severely affect health systems. A number of countries in LMIC are considered as fragile countries and states (FCAS) due to ongoing wars and or chronic political instability rendering these countries as low priority for research (Bisika 2010; Kamuzora et al. 2013; Woodward et al. 2017). Political differences and instability are a source of wars and conflicts around the world. The impacts of political upheavals span the whole range of mental, social, and physical wellbeing. Apart from the physical trauma, the social impacts and impacts of mental health are often underestimated. With existing strained health systems in some if not most of LMIC, the complications brought about by political instability further overwhelm and often result in the collapse of the already fragile health system (Bürgin et al. 1974, 2022; Punamäki et al. 2015; Slone et al. 2017).

An online survey comprising of respondents from various backgrounds and setting revealed several factors that contribute to poor health systems research (HSR) in FCAS (Woodward et al. 2017). These include difficulty to attract and convince stakeholders to invest in research in such fragile environments including post-conflict situations. Some of the key stakeholders are national governments, international donors, and academic institutions. Some of the National governments in FCAS are not economically stable to manage the basic needs of their populations and research may not be a priority. The lack of interest from stakeholders ushers in the challenge of funding, while there is disease specific research that maybe prioritized, HSR is often not in that category. On the other hand there are times that funding may be made available but most often this funding is targeted and often short-lived resulting in lack of continuity and sustainability. With minimal funding and poor sustainability, HSR in LMIC especially FCAS remains to be attained (Woodward et al. 2017).

Conflict has resulted in forced migration and displacement for many. The impact of wars, forced migration and displacement, is immense affecting both children and adults, with their health and wellbeing severely disturbed (Burkle Jr 2017; Kevlihan 2013; Hotez et al. 2012). The displaced populations have minimal or no access to healthcare severely affecting children's health and development as well as maternal health. While children miss out on the required vaccines for vaccine preventable diseases, adults face the challenges of lack of diagnosis for chronic conditions including cancers, hypertension, diabetes, and other age related diseases (Dhavan et al. 2017; World Health Organization - Regional Office for Europe 2017). The poor living conditions of displaced populations expose the affected individuals to infectious diseases and poor nutrition renders these populations vulnerable to disease (Bürgin et al. 1974, 2022; Punamäki et al. 2015).

Brain Drain

Public health innovation refers to development of new tools that enhance efficiency, productivity and impact to improve health outcomes through various expanses including policies, products, equipment and or software and many more. There is need for innovations that are applicable and timely for LMIC. Most of the current innovations are quite costly and highly unlikely to be sustained in limited resource settings. Limited access to information, tools and opportunities in LMIC has led to a phenomenon known as "Brain Drain". Researchers are often keen to develop new tools and pioneer innovations. With brain drain, highly qualified individuals such as medical practitioners and researchers move to countries with better economies. Frustration has led to brain drain forcing individuals to seek greener pastures. For example, a highly trained health practitioner may face a preventable situation such as loss of life due to lack of necessary resources to diagnose, treat and or manage a treatable disease. Similarly, a highly qualified researcher may not be able to carry out advanced research in their own country due to lack of facilities and equipment. This has resulted in highly competent researchers opting to work in countries with ample resources, hence brain drain. LMIC face the challenge of lack of equipment and or resources which could contribute to the frustration and brain drain. (Ekenze et al. 2014).

Quote 2.1

Imagine two individuals studying medicine in different settings. While one has access to all the technology to make a clear diagnosis of a problem, one is struggling to achieve the same diagnosis but without the associated technology. While a patient is monitored at every second during the process of attaining a diagnosis in one setting, on the other side a patient is left lying in agony while the health practitioner is trying to figure out where they can find a dose of medicine that can stabilize the patient; and have no way of achieving a diagnosis than elimination methods through evaluation of which treatment is working or not.

2.2 Identifying the Gaps

> I took a visit to one of the innovation centres at one university campus. It was an occasion where they were show casing various innovations and how they can be used in studying medicine; simultaneously encouraging creativity and innovation mindset in medical students by identifying gaps of knowledge and how these gaps can be addressed. This got me thinking can such a mindset be achievable in LMIC? How can health practitioners and public health experts in this region tap into this knowledge?

Capacity building in LMIC has been a discussion point for the past decades to help address the knowledge gap as well as enhance skills in the local population. Various global health and public health programs have been established in various institutions around the world, some of which offer bursaries for people from LMIC. While the knowledge gap is being addressed there is another gap that has been widening between HIC and LMIC and this is the technology gap. In as much as there are skilled health practitioners in LMIC, the limited technology and infrastructure impedes on their capacity to provide effective healthcare services to their respective populations. This outcome shows that addressing the healthcare needs in LMIC requires effective operation not only of one sector but various sectors which directly and indirectly determine health outcomes. For example, a poorly functioning health system does not imply failure in the healthcare sector only but also various sectors including economic and social sectors. Success of a healthcare system relies both on the individual as well as the functional aspects of the system itself. A concept of integrated healthcare has been under discussion in the past decade and is gradually being adopted in some healthcare settings around the world. A review by Suter et al., outlined ten points to be considered in integrated healthcare (Suter et al. 2009).

Identifying gaps of knowledge and finding ways to address the need is one key factor to achieve a viable innovation. Innovations that are timely, affordable, and able to meet required needs in each population are highly likely to be well accepted within that population than those that do not. Various settings come with their own needs and addressing the needs require a closer look at the end user and how these innovations will benefit them. Since the dawn of the new millennium, there has been an increase in innovations in medical technology to tackle the various needs. Despite these achievements the rate at which these innovations are advancing are not the same in different settings. While high income countries are becoming more and more advanced with new therapies and devices, LMIC are often left behind not being able to attain the same advancements (Kruk et al. 2018; Ngoc Dinh et al. 2020). Despite some of these technologies and therapies being readily available, access to these is a challenge in most populations in LMIC. The cost of purchasing and maintaining most of these technologies is not attainable in most LMIC.

Innovations require not only understanding the gaps, but also having the technical know-how to attain the required innovation. Innovations require multidisciplinary approaches often including computer science, artificial intelligence, and ergonomics. Understanding people in their environments is one way to understand the needs as well as approaches in which these needs can be met. While one may be able to understand ergonomics, computer science and artificial intelligence are fields that require a great deal of technical know-how and infrastructure. In the current state, most LMIC are lagging in their ability to generate enough power as well as finances to manage advance technological studies. Innovations require both financial and human resources including willing sponsors, such as governments that can sustain and advance the changes.

Design thinking revolves around several core concepts including exploring the need, empathy, envisioning how these needs will be attained, experimentation and execution. Each of these steps requires not only manpower but also financial commitment every step of the way until the product is delivered to the end user (Bazzano et al. 2017).

In global health, innovations are needed to meet the various needs of populations in various settings around the world. Understanding the social needs, structures and challenges in various populations is necessary to attain innovations that are user friendly within that population. The challenge nowadays is exclusion of certain populations in the ideation process (Bangert et al. 2017; Bedriñana et al. 2021; Guzel et al. 2021). In as much as an innovation might be relevant in one population, it may not serve the designated purpose as required in another setting and population.

How Can LMIC Attain and Sustain Innovation Within Their Settings?

While most governments in LMIC are overwhelmed financially, private sector is often vibrant and often having much financial prowess. Thus, there is need for multisectoral collaboration either between governments and private sector as well as within the private sector to finance and sponsor research and innovation within their countries (Ehrenberg et al. 2020; Kuruvilla et al. 2018; Nkengasong et al. 2017). This approach will not only allow for innovations that meet the needs of the population within the country but also improve the health and wellbeing of these populations. With most of the LMIC being hosts to infectious diseases and neglected tropical diseases, innovations originating from these countries could assist in solving these issues (Bangert et al. 2017). While most of the endemic diseases might not affect other populations outside endemic countries, it is highly unlikely that non-endemic areas will invest in innovations to solve challenges that do not directly affect their populations. It is therefore important for endemic regions and countries to focus more on managing the needs of their populations through targeted innovations that are affordable and sustainable within their settings.

Rural populations as well as populations living it utmost poverty are often neglected in policies as such these populations are often left behind and are underserved. Consequently, it is these populations that are often plagued with infectious diseases, including neglected diseases, emerging, and re-emerging diseases. Management of chronic non communicable diseases within these populations is another challenge e.g., management of diabetes, access and cost of interventions are often out of reach for such populations (Mediannikov et al. n.d.; Nordin et al. 2017; Zafar and McQueen 2011).

2.2.1 Integrated Healthcare System

Integrated healthcare is an approach whose main goal is to provide a holistic treatment by combining physical, mental, behavioral and financial aspects of healthcare. This collaborative approach focusses on the patient and collaboration across wide-ranging services. Although this approach is mainly focused on management of chronic diseases, it is imperative that it should be considered in discussions of infectious disease. Some infectious diseases could lead to both physical debilitation, social exclusion and can seriously affect the mental health of infected individuals. Similarly, some chronic diseases could lead to an individual being at risk of certain infectious diseases. For example, diabetes, and hypertension are some conditions that could put an individual at risk of infectious diseases such as COVID-19 and TB. Successful management of these chronic diseases as well as mental health could reduce this risk to a certain extent (Bosire et al. 2021; Suter et al. 2009).

According to Suter et al. the following are the ten points to consider in integrated healthcare approach (Table 2.1).

2.2.1.1 Cooperation Between Health and Social Care Organizations

Multi sector cooperation can be applied in public health preparedness and response and is a crucial component in attaining sustainable development goals (Kuruvilla et al. 2018). In Global health, multi sectoral approach is the key strategy for Women's, Children's and Adolescents' Health and is essential in global health priorities including universal health coverage, prevention and control of communicable disease as well as prevention and control of non-communicable diseases (Manual 2017; UN 2015; WHO 2018; World Health Organization (WHO) 2017). Attaining multi sectoral collaboration requires commitment from stakeholders these include individuals or groups, governments, nongovernmental organizations (NGOs), civil society, private sector, international organizations, donors, service users and providers, the media, and other groups.

Multisectoral collaboration can be attained both locally and at international level. At the local level, such collaboration could assist in planning, budgeting, financing, management, implementation as well as monitoring and evaluation to achieve desired outcomes (Kuruvilla et al. 2018). Such collaboration could determine success or failure of desired health outcomes depending on the effectiveness of the collaboration. With proper planning, contributions from individual sectors working independently could assist in advancing goals in other sectors. For

Table 2.1 Ten points to consider in integrated healthcare approach (Source: Suter et al. 2009; Suter et al. 2009)

Approach	Focus area
Cooperation between health and social care organizations	• Access to care continuum with multiple points of access • Emphasis on wellness, health promotion and primary care
Patient focus	• Patient-centreed philosophy; focusing on patients' needs • Patient engagement and participation • Population-based needs assessment; focus on defined population
Geographic coverage and rostering	• Maximize patient accessibility and minimize duplication of services • Roster: responsibility for identified population; right of patient to choose and exit
Standardized care delivery through interprofessional teams	• Interprofessional teams across the continuum of care • Provider-developed, evidence-based care guidelines and protocols to enforce one standard of care, regardless of where patients are treated
Performance management	• Committed to quality of services, evaluation and continuous care improvement • Diagnosis, treatment, and care interventions linked to clinical outcomes
Information systems	• State of the art information systems to collect, track and report activities • Efficient information systems that enhance communication and information flow across the continuum of care
Organizational culture and leadership	• Organizational support with demonstration of commitment • Leaders with vision who are able to instill a strong, cohesive culture
Physician integration	• Physicians are the gateway to integrated healthcare delivery systems • Pivotal in the creation and maintenance of the single-point-of-entry or universal electronic patient record • Engage physicians in leading role, participation on board to promote buy-in
Governance structure	•Strong, focused, diverse governance represented by a comprehensive membership from all stakeholder groups • Organizational structure that promotes coordination across settings and levels of care
Financial management	• Aligning service funding to ensure equitable funding distribution for different services or levels of services • Funding mechanisms must promote interprofessional teamwork and health promotion • Sufficient funding to ensure adequate resources for sustainable change

example determinants of health are not only health related but also social, economic and environmental causes (Mphande 2016). As such addressing the social, economic, and environmental challenges in the community and society could determine healthier outcomes and greater chances of attaining SDG3 and the rest of the SDGs.

At Intentional level, collaboration between various stake holders to attain common aspects including provision of resources e.g. joint programmes, donations in form of equipment and or financial support, skills exchange, data sharing, collaborative surveillance goals are crucial. To attain such collaborations, there is need for trust between the various stakeholders to ensure that they are all working together towards a common cause. International collaboration could involve coming together and collaborating in a managed process to achieve shared outcomes. Previous disease outbreaks such as the Ebola outbreaks, and the COVID-19 pandemic have emphasized the need for a collaborated and coordinated response between various sectors to combat infectious disease outbreaks. While the cooperation between health and social organizations has been implemented successfully in some countries, most LMIC still have a long way to achieve this (De Leeuw 2017; Jayatilleke 2020; Levinson et al. 2014; Wang et al. 2014). Successes of international multisectoral collaborations have been observed in many parts of the world including LMIC. It was observed that during the millennium development goals (between 2000 and 2015), health outcomes of children and women were improved with approximately 50% reduction in child and maternal mortality. Improvement in health development and other sectors were also observed (Shi et al. 2017; Wang et al. 2014).

2.2.1.2 Patient Focus

Apart from collaboration, focus on the needs of a patient is another aspect of achieving integrated health systems. Different patients have varying needs as such focusing on meeting these needs is essential. For better health outcomes, allowing patient engagement and participation to achieve health needs could assist in bringing desirable change and outcomes both for the patient and the health system. Poor engagement gives room to non- compliance and poor health outcomes. Understanding the needs of different populations is another aspect not to be overlooked. Different populations have varying needs depending on their genetic disposition, setting, and economic prowess. As such needs assessment is required to ensure that each population is served accordingly and that set goals are met with each population. Patient centred care approach has been implemented in some countries in LMIC (Haroun et al. 2022).

In South Africa, integrated chronic disease management (ICDM) which has patient centred care (PCC) component has been implemented. Despite the success in implementations, several challenges were observed, these included systemic challenges and patient barriers. Staff shortages, lack of guidelines for comorbid care, and fragmented care, were some of the systemic challenges observed. Similarly patient barriers such as language, poverty, missed appointments were also reported. Despite the challenges the objective of ICDM was to empower patients with HIV and Diabetes (chronic disease) to play an active role in the disease management process and intervention at a community/population and health service level (Bosire et al. 2021).

In Tanzania, the Afya-Tek digital intervention, aimed at bridging the knowledge gaps on the burden of disease and related health and technology. The study looked at health seeking behaviour; workflow procedures and challenges experiencing healthcare actors; adolescent health and health seeking behaviour; and technological literacy and perceptions on the use of digital technologies in healthcare delivery. The goal of Afya-Tek digital intervention was to create a product that is patient centred to carter for the needs of the patients. Although the intervention is still in its infancy, a formative research study showed the benefits of engaging users during the development of the technology (Haroun et al. 2022). Below are some case studies to illustrate PCC and its advantages.

2.2.2 Case Study 1: South Africa

2.2.2.1 Patient-Centred Care for Patients with Diabetes and HIV at a Public Tertiary Hospital in South Africa: An Ethnographic Study (Bosire et al. 2021)

Edna N. BosireEmily Mendenhall, Shane A. Norris and Jane Goudge

Keywords Diabetes, HIV/AIDS, Patient-Centred Care, South Africa

Study Aims

The study was conducted in South Africa, a country in Southern Africa. South Africa has implemented an integrated chronic disease management (ICDM) which has PCC component. The program was aimed at empowering HIV and Diabetes chronic care patients in taking up an active role in the disease management process as well as bring intervention at a community/population and health service level. Systemic and structural challenges were reported affecting implementation of PCC. These included staff shortages, lack of guidelines for comorbid care, and fragmented care, and patient barriers such as poverty, language, and missed appointments. The paper concluded that health systems need to be strengthened and made the following recommendations.

1. *Multidisciplinary guidelines*: Ensuring appropriate multidisciplinary guidelines for managing comorbidities exist, are known, and available
2. *Strengthening primary healthcare*: Strengthening primary healthcare (PHC) clinics by ensuring access to necessary resources that will facilitate successful integration and management of comorbid diabetes and HIV
3. *Training:* Training medical practitioners on PCC and structural competence, so as to better understand patients in their sociocultural contexts
4. *Understanding patient challenges*: Understanding patient challenges to effective care to improve attendance and adherence.

Key Message

A. Implications for Policy Makers
   - Emphasis on the need to strengthen health systems in South Africa and other similar contexts
     – by increasing the number of staff and developing multidisciplinary guidelines for managing comorbidities. Which would also facilitate implementation of integrated and PCC.
   - Policymakers can improve care for people with comorbid diabetes and HIV by ensuring sufficient equipment and trained staff at primary healthcare (PHC) clinics in South Africa.
   - Medical institutions need to invest in training and equipping medical practitioners and students with cultural humility and structural competence skills that would help them adopt PCC.

B. Implications for the Public
   - Need for a close relationship between healthcare providers and patients in terms of chronic disease management.
   - Patients' socio-cultural and economic backgrounds cannot be ignored within clinical spaces.
   - Patients' voices, perspectives and lived experiences of chronic diseases are important especially when designing chronic care programs that are contextually appropriate.
   - Providers need to support patients with and often also, their informal carers' in improving their knowledge and skills and encouraging them to actively participate and collaborate with providers in decision-making and self-management.

2.2.3 Case Study 2: Tanzania

2.2.3.1 A Human Centred Approach to Digital Technologies in Health Care Delivery Among Mothers, Children and Adolescents (Haroun et al. 2022)

Yasini HarounRichard Sambaiga,
Nandini SarkarNtuli A. Kapologwe
James KengiaJafary LianaSuleiman Kimatta
Johanita JamesVendelin Simon
Fatma HassanRomuald Mbwasi
Khadija FumbweRebecca Litner
Gloria Kahamba and Angel Dillip

Study Aims

The formative research aimed to examine

1. *The burden of disease and related health seeking behaviour*
2. *Workflow procedures and challenges experiencing healthcare actors*
3. *Adolescent health and health seeking behaviour*
4. *Technological literacy and perceptions on the use of digital technologies in healthcare delivery.*

The study was conducted in Tanzania, a country in East Africa. Healthcare outcomes in child, adolescent and maternal in Tanzania are poor, and mostly characterized by fragmentary service provision. An initiative known as Afya-Tek aimed at connecting various stakeholders was planned to be implemented. Prior to implementation this study was conducted. Afya-Tek is a digitally enabled, community based responsive health system initiative that brings together the key health actors at the community level. The project linked Community Health Volunteers (CHWs), Accredited Drug Dispensing Outlets (ADDOs), and public health facilities with the aim of improving decision making, prompt access to care through referrals and overall quality of care along the continuum of care. This would be achieved by using the digital platform recommended by the Government of Tanzania (Open Smart Register Platform), where these key players will be empowered to efficiently refer patients to the appropriate service, track their status, and follow up to ensure proper and respectful care. The Afya-Tek initiative was guided by three approaches: (a) human-centred participatory design necessary to inform the design of the Afya-Tek digital intervention; (b) digital health; and (c) biometric identification.

Key Findings and Conclusions

The study found that there were multiple work-related challenges including

- Lack of proper mechanism to track referrals and patient's information were noted across healthcare actors.
- There was a keen interest in the use of technologies shown by all study participants to improve care coordination and health outcomes among health system actors.
- Participants shared their views on how they envision the digital system working.

The findings from the formative study were used for informing the subsequent phases of the co-development and implementation of the Afya-Tek digital health intervention; with a view to making it relevant to the needs of those who will use it in the future. The findings envisaged the best ways to design digital intervention tailored to meet the needs of those who will be using it.

From the two case studies, there is evidence that with the availability of resources a PCC approach is possible, and this approach would also help in identifying gaps and improve healthcare delivery.

2.3 Neglected Diseases

Apart from identifying gaps, there are some areas that have been identified and action is required so that "no-one is left behind" and this is the area of neglected tropical diseases (NTD). Neglected tropical diseases affect the most vulnerable and invisible populations of the world (Bangert et al. 2017). These populations and infections are invisible in that they affect most vulnerable populations who are often disconnected from parts of society due to poverty, debilitating diseases, stigma, lack of access to primary healthcare and facilities and various social economic factors (Bedriñana et al. 2021). Research and campaigns aimed at reducing and combating NTD in the affected populations has made strides during the past decade. The launch of the WHO road map for NTD entitled 'Ending the neglect to attain the Sustainable Development Goals' paves way to eliminating NTD (World Health Organisation (WHO) 2016). With the roadmap pushing for Stronger accountability, intensified cross-cutting approaches, and a change in operating model and culture how does this roadmap make a difference in LMIC? There is need for commitment from countries to ensure that the roadmap is implemented and followed (Glenn et al. 2021; Souza et al. 2021). What is the future for fight against NTD and can elimination of NTD be sustained in the wake of socioeconomic disparities in LMIC? In the last decade, several NTD such as Trachoma, Onchocerciasis, and River blindness have been successfully eliminated in some countries (Nicholls et al. 2018). While these success stories are celebrated there is growing concerns of sustainability. Will the countries that have successfully eliminated some of the NTD be able to sustain this status? A study conducted in Kenya reported some of the factors that could influence the success of NTD sustainability programs (Thuo et al. 2020). These include, programmatic factors, contextual factors and technical factors. Programmatic factors include leadership, capacity, flexibility, integration, and performance of the program. While contextual factors focus on political commitment, reliable funding, political stability, and environmental improvement. Technical factors focus more on treatment and control, these include drug type, mass drug administration (MDA), vector control and morbidity management. Comparing programmatic and contextual factors, the later are more relevant as these determine the environment in which the programme will be implemented. Another factor that could influence sustainability of NTD programmes is availability of reliable funding, which is a major contextual factor (Kollie et al. 2020). The study also identified new factors which have significant influence on sustainability, these include government systems, rural-urban environment, education level, cultural beliefs, and lifestyle. The role of the community was identified as a new programmatic factor; though not highly ranked in relevance (Thuo et al. 2020). While countries are moving at different paces to achieve the NTD elimination goals effective mapping and intervention strategies in NTD endemic countries would help to identify and address challenges both in countries facing complex emergencies such as conflict and those without (Kelly-Hope et al. n.d.). Evaluation of current strategies and their challenges and identifying locally relevant strategies in endemic countries could be one approach to achieve the NTD-2030 roadmap (Onasanya et al. 2021). Neglected tropical disease modelling has assisted in making forecasts towards achieving NTD elimination, such strategies are crucial for policy makers for informed decisions (Clark et al. 2021). The role of pharmaceutical companies that have donate the medicines that are used in mass drug administration (MDA) for the past decades have enabled NTD elimination programs globally, as such sustainability and achievement of NTD elimination rests on these partnerships (Bradley et al. 2021).

2.4 Conclusion

There is no sustainable health if there are still communities that have no access to healthcare, if there are still neglected diseases, if there are still vaccine inequities, if there are health inequalities and systematic exclusion of the most vulnerable and affected populations in decision making for their health and wellbeing. The response to the COVID-19 pandemic has shown that countries are willing to leave others behind as long as they keep out populations that are the "source" of the disease. These polices are made to suit part of the global population and not for all. These decisions are made to affect everyone globally with minimal and or complete lack of consideration for LMIC. The "bread crumb" policy seen during the COVID-19 pandemic where the LMIC must wait for leftover vaccines from high income countries are testament that there are still billions globally that are still left behind.

References

Abu-Odah H, Su J, Musa SS (2022) Unwillingness or reluctance of Palestinians to get the COVID-19 vaccine: the reasons behind it and how to persuade them. Int J Infect Dis 119:53. https://doi.org/10.1016/J.IJID.2022.03.039

Addo-Atuah J, Senhaji-Tomza B, Ray D, Basu P, Loh FHE, Owusu-Daaku F (2020) Global health research partnerships in the context of the sustainable development goals (SDGs). Res Soc Adm Pharm 16(11):1614–1618. https://doi.org/10.1016/j.sapharm.2020.08.015

Afsahi S, Lerner MB, Goldstein JM, Lee J, Tang X, Bagarozzi DAJ et al (2018) Novel graphene-based biosensor for early detection of Zika virus infection. Biosens Bioelectron 100:85–88. https://doi.org/10.1016/j.bios.2017.08.051

Ahmed Ali H, Hartner AM, Echeverria-Londono S, Roth J, Li X, Abbas K et al (2022) Correction: vaccine equity in low and middle income countries: a systematic review and meta-analysis (International Journal for Equity in Health (2022); 21(82), 10.1186/s12939-022-01678-5). Int J Equity Health 21(1):1–30. https://doi.org/10.1186/s12939-022-01695-4

Arce JSS, Warren SS, Meriggi NF, Scacco A, Mcmurry N, Voors M et al (2021) COVID-19 vaccine acceptance and hesitancy in low- and middle-income countries. Nat Med 27:1385–1394. https://doi.org/10.1038/s41591-021-01454-y

Bangert M, Molyneux DH, Lindsay SW, Fitzpatrick C, Engels D (2017) The cross-cutting contribution of the end of neglected tropical diseases to the sustainable development goals. Infect Dis Poverty 6(1):73. https://doi.org/10.1186/s40249-017-0288-0

Barrett MP, Croft SL (2012) Management of trypanosomiasis and leishmaniasis. Br Med Bull 104(1):175–196. https://doi.org/10.1093/bmb/lds031

Bazzano AN, Martin J, Hicks E, Faughnan M, Murphy L (2017) Human-centred design in global health: a scoping review of applications and contexts. PLoS One 12(11):e0186744. https://doi.org/10.1371/journal.pone.0186744

Bedriñana KGA, Martín JAR, Añaños FT (2021) Human rights in the least developed countries of Asia: an index for quantifying sustainable development goal 3 (good health and wellbeing). Int J Environ Res Public Health 18(9):4747. https://doi.org/10.3390/ijerph18094747

Bisika T (2010) Health systems strengthening in conflict situations. East Afr J Public Health 7(3):277–281

Bosire EN, Mendenhall E, Norris SA, Goudge J (2021) Patient-centred care for patients with diabetes and hiv at a public tertiary hospital in South Africa: an ethnographic study. Int J Health Policy Manag 10(9):534–545. https://doi.org/10.34172/ijhpm.2020.65

Bradley M, Taylor R, Jacobson J, Guex M, Hopkins A, Jensen J et al (2021) Medicine donation programmes supporting the global drive to end the burden of neglected tropical diseases. Trans R Soc Trop Med Hyg 115:136–144. https://doi.org/10.1093/trstmh/traa167

Bürgin D, Anagnostopoulos D, Vitiello B, Sukale T, Schmid M, Fegert JM (1974) Impact of war and forced displacement on children's mental health-multilevel, needs-oriented, and trauma-informed approaches. Eur Child Adolesc Psychiatry 31:845–853. https://doi.org/10.1007/s00787-022-01974-z

Bürgin D, Anagnostopoulos D, Doyle M, Eliez S, Fegert J, Fuentes J et al (2022) Impact of war and forced displacement on children's mental health—multilevel, needs-oriented, and trauma-informed approaches. Eur Child Adolesc Psychiatry 31(6):845–853. https://doi.org/10.1007/S00787-022-01974-Z

Burkle FM Jr (2017) The politics of global public health in fragile states and ungoverned territories. PLoS Currents 9:ecurrents.dis.ba3beede71ca0746a0972aa3837ed618. https://doi.org/10.1371/currents.dis.ba3beede71ca0746a0972aa3837ed618

CDC (2022) Treatment information for healthcare professionals. U.S. Department of Health & Human Services. https://www.cdc.gov/poxvirus/monkeypox/clinicians/treatment.html. Accessed 7 Feb 2023

Chaccour J (2018) Authorship trends in the lancet global health: only the tip of the iceberg? Lancet Glob Health 6(5):e497. https://doi.org/10.1016/S2214-109X(18)30110-4

Clark J, Stolk WA, Basáñez M-G, Coffeng LE, Cucunubá ZM, Dixon MA et al (2021) How modelling can help steer the course set by the World Health Organization 2021–2030 roadmap on neglected tropical diseases. Gates Open Res 5:112. https://doi.org/10.12688/gatesopenres.13327.1

De Leeuw E (2017) Nature and health. Annu Rev Public Health 38:179. https://doi.org/10.1146/annurev-publhealth

Deml MJ, Githaiga JN (2022) Original research: determinants of COVID-19 vaccine hesitancy and uptake in sub-Saharan Africa: a scoping review. BMJ Open 12(11):e066615. https://doi.org/10.1136/BMJOPEN-2022-066615

Dhavan P, Dias HM, Creswell J, Weil D (2017) An overview of tuberculosis and migration. Int J Tuberc Lung Dis 21(6):610–623. https://doi.org/10.5588/ijtld.16.0917

Eccleston-Turner M, Upton H (2021) International collaboration to ensure equitable access to vaccines for COVID-19: the ACT-accelerator and the COVAX facility. Milbank Q 99(2):426–449. https://doi.org/10.1111/1468-0009.12503

Ehrenberg JP, Zhou X-N, Fontes G, Rocha EMM, Tanner M, Utzinger J (2020) Strategies supporting the prevention and control of neglected tropical diseases during and beyond the COVID-19 pandemic. Infect Dis Poverty 9:86. https://doi.org/10.1186/s40249-020-00701-7

Ekenze SO, Onumaegbu OO, Nwankwo OE (2014) The current status of international partnerships for child surgery in sub-saharan Africa. Int Surg 99:616–622. https://doi.org/10.9738/INTSURG-D-13-00244.1

Fang E, Liu X, Li M, Zhang Z, Song L, Zhu B et al (2022) Advances in COVID-19 mRNA vaccine development. Signal Transduct Target Ther 7:94. https://doi.org/10.1038/s41392-022-00950-y

Ghani M, Hurrell R, Verceles AC, McCurdy MT, Papali A (2020) Geographic, subject, and authorship trends among LMIC-based scientific publications in high-impact Global Health and general medicine journals: a 30-month Bibliometric analysis. J Epidemiol Glob Health 11(1):92. https://doi.org/10.2991/jegh.k.200325.001

Glenn J, Adams AB, Sankar G, Henry C, Palacio K, Thuo W, Williams K (2021) Towards a shared understanding of sustainability for neglected tropical disease programs. PLoS Negl Trop Dis 15(8):1–12. https://doi.org/10.1371/journal.pntd.0009595

Guzel AE, Arslan U, Acaravci A (2021) The impact of economic, social, and political globalization and democracy on life expectancy in low-income countries: are sustainable development goals contradictory? Environ Dev Sustain 23(9):13508–13525. https://doi.org/10.1007/s10668-021-01225-2

Haroun Y, Sambaiga R, Sarkar N, Kapologwe NA, Kengia J, Liana J et al (2022) A human centred approach to digital technologies in health care delivery among mothers, children and adolescents. BMC Health Serv Res 22(1):1–15. https://doi.org/10.1186/s12913-022-08744-2

Hasan BS, Rasheed MA, Wahid A, Kumar RK, Zuhlke L (2021) Generating evidence from contextual clinical research in low- to middle income countries: a

roadmap based on theory of change. Front Pediatr 9:764239. https://doi.org/10.3389/FPED.2021.764239
Hotez PJ, Ferris MT (2006) The antipoverty vaccines. Vaccine 24(31–32):5787–5799. https://doi.org/10.1016/j.vaccine.2006.05.008
Hotez PJ, Savioli L, Fenwick A (2012) Neglected tropical diseases of the middle east and north africa: review of their prevalence, distribution, and opportunities for control. PLoS Negl Trop Dis 6(2):e1475. https://doi.org/10.1371/journal.pntd.0001475
Jayatilleke K (2020) Challenges in implementing surveillance tools of high-income countries (HICs) in low middle income countries (LMICs). Curr Treat Options Infect Dis 12(3):191–201. https://doi.org/10.1007/s40506-020-00229-2
Kamuzora P, Maluka S, Ndawi B, Byskov J, Hurtig AK (2013) Promoting community participation in priority setting in district health systems: experiences from Mbarali district, Tanzania. *Glob Health Action* 6:22669. https://doi.org/10.3402/gha.v6i0.22669
Kelaher M, Ng L, Knight K, Rahadi A (2016) Equity in global health research in the new millennium: trends in first-authorship for randomized controlled trials among low- and middle-income country researchers 1990-2013. Int J Epidemiol 45(6):2174–2183. https://doi.org/10.1093/ije/dyw313
Kelly-Hope LA, Sanders AM, Harding-Esch E, Willems J, Ahmed F, Vincer F, Hill R (n.d.) Complex emergencies and the control and elimination of neglected tropical diseases in Africa: developing a practical approach for implementing safe and effective mapping and intervention strategies. Confl Heal 15(1):18. https://doi.org/10.1186/s13031-021-00356-7
Kevlihan R (2013) Providing health services during a civil war: the experience of a garrison town in South Sudan. Disasters 37(4):579–603. https://doi.org/10.1111/disa.12027
Kollie K, Siakeh A, Zawolo G, Wickenden A, Theobald S, Rogers E et al (2020) Donor reliance and the impact on neglected tropical disease programme delivery: reflections and solutions for change from programme management perspectives. Int Health 0:1–3. https://doi.org/10.1093/inthealth/ihaa090
Kruk ME, Gage AD, Joseph NT, Danaei G, García-Saisó S, Salomon JA (2018) Mortality due to low-quality health systems in the universal health coverage era: a systematic analysis of amenable deaths in 137 countries. Lancet 392(10160):2203–2212. https://doi.org/10.1016/S0140-6736(18)31668-4
Kuruvilla S, Hinton R, Boerma T, Bunney R, Casamitjana N, Cortez R et al (2018) Business not as usual: how multisectoral collaboration can promote transformative change for health and sustainable development. BMJ 363:k4771. https://doi.org/10.1136/bmj.k4771
Leclerc H, Schwartzbrod L, Dei-Cas E (2002) Microbial agents associated with waterborne diseases. Crit Rev Microbiol 28(4):371–409. https://doi.org/10.1080/1040-840291046768
Lee M, You M (2022) Direct and indirect associations of media use with COVID-19 vaccine hesitancy in South Korea: cross-sectional web-based survey. J Med Internet Res 24(1):e32329. https://doi.org/10.2196/32329
Levinson KL, Abuelo C, Chyung E, Belinson SE, Sologuren CV, Ortiz CS et al (2014) The Peru cervical cancer prevention study (PERCAPS): community based participatory research in Manchay, Peru. Int J Gynecol Cancer 23(1):141–147. https://doi.org/10.1097/IGC.0b013e318275b007.The
Luna N, Ramírez AL, Muñoz M, Ballesteros N, Patiño LH, Castañeda SA et al (2022) Phylogenomic analysis of the monkeypox virus (MPXV) 2022 outbreak: emergence of a novel viral lineage? Travel Med Infect Dis 49:102402. https://doi.org/10.1016/J.TMAID.2022.102402
Manual T (2017) WHO health in all policies training manual. WHO. http://www.who.int/social_determinants/publications/health-policies-manual/en/
Mediannikov O, Diatta G, Fenollar F, Sokhna C, Trape J-F, Raoult D (n.d.) Tick-borne Rickettsioses, neglected emerging diseases in rural Senegal. PLoS Negl Trop Dis 4(9):e821. https://doi.org/10.1371/journal.pntd.0000821
Mphande FA (2016) Infectious diseases and rural livelihood in developing countries. Springer. https://doi.org/10.1007/978-981-10-0428-5
Nass SJ, Levit LA, Gostin LO (2009) Beyond the HIPAA privacy rule: enhancing privacy, improving health through research. National Academies Press, pp 1–320. https://doi.org/10.17226/12458
Newton CR (2020) Research and open access from low- and middle-income countries. Dev Med Child Neurol 62(5):537–537. https://doi.org/10.1111/dmcn.14513
Ngoc Dinh M, Nygate J, Hoang Minh Tu V, Thwaites CL (2020) New technologies to improve healthcare in low- and middle-income countries: global grand challenges satellite event, Oxford university clinical research unit, Ho Chi Minh City, 17th-18th September 2019. Wellcome Open Res 5:142. https://doi.org/10.12688/wellcomeopenres.16008.2
Nicholls RS, Duque S, Olaya LA, López MC, Sánchez SB, Morales AL, Palma GI (2018) Elimination of onchocerciasis from Colombia: first proof of concept of river blindness elimination in the world. Parasit Vectors 11(1):237. https://doi.org/10.1186/s13071-018-2821-9
Nkengasong J, Djoudalbaye B, Maiyegun O (2017) A new public health order for Africa's health security. Lancet Glob Health 5:e1064–e1065. https://doi.org/10.1016/S2214-109X(17)30363-7
Nordin P, Thielecke M, Ngomi N, Mudanga GM, Krantz I, Feldmeier H (2017) Treatment of tungiasis with a two-component dimeticone: a comparison between moistening the whole foot and directly targeting the embedded sand fleas. Trop Med Health 45(1):6. https://doi.org/10.1186/s41182-017-0046-9
Onasanya A, Bengtson M, Oladepo O, Van Engelen J, Diehl JC (2021) Rethinking the top-down approach to Schistosomiasis control and elimination in sub-

Saharan Africa. Front Public Health 9:622809. https://doi.org/10.3389/fpubh.2021.622809
Pai M (n.d.) Passport and visa privileges in global health.. https://www.forbes.com/sites/madhukarpai/2022/06/06/passport-and-visa-privileges-in-global-health/?sh=444be6034272. Accessed 18 Mar 2023
Peacocke EF, Heupink LF, Frønsdal K, Dahl EH, Chola L (2021) Global access to COVID-19 vaccines: a scoping review of factors that may influence equitable access for low and middle-income countries. BMJ Open 11:49505. https://doi.org/10.1136/bmjopen-2021-049505
Punamäki RL, Palosaari E, Diab M, Peltonen K, Qouta SR (2015) Trajectories of posttraumatic stress symptoms (PTSS) after major war among Palestinian children: trauma, family- and child-related predictors. J Affect Disord 172:133–140. https://doi.org/10.1016/j.jad.2014.09.021
Quick J, Loman NJ, Duraffour S, Simpson JT, Severi E, Cowley L et al (2016) Real-time, portable genome sequencing for Ebola surveillance. Nature 530(7589):228–232. https://doi.org/10.1038/nature16996
Rosa WE, Ahmed E, Chaila MJ, Chansa A, Cordoba MA, Dowla R et al (2022) Can You hear us now? Equity in global advocacy for palliative care. J Pain Symptom Manag 64(4):e217–e226. https://doi.org/10.1016/j.jpainsymman.2022.07.004
Schnee M, Vogel AB, Voss D, Petsch B, Baumhof P, Kramps T, Stitz L (2016) An mRNA vaccine encoding rabies virus glycoprotein induces protection against lethal infection in mice and correlates of protection in adult and Newborn pigs. PLoS Negl Trop Dis 10(6):e0004746. https://doi.org/10.1371/journal.pntd.0004746
Seimenis A (2012) Zoonoses and poverty - a long road to the alleviation of suffering. Vet Ital 48(1):5–13
Shi T, McAllister DA, O'Brien KL, Simoes EAF, Madhi SA, Gessner BD et al (2017) Global, regional, and national disease burden estimates of acute lower respiratory infections due to respiratory syncytial virus in young children in 2015: a systematic review and modelling study. Lancet 390(10098):946–958. https://doi.org/10.1016/S0140-6736(17)30938-8
Slone M, Lavi I, Ozer EJ, Pollak A (2017) The Israeli-Palestinian conflict: meta-analysis of exposure and outcome relations for children of the region. Child Youth Serv Rev 74:50–61. https://doi.org/10.1016/j.childyouth.2017.01.019
Souza AA, Ducker C, Argaw D, King JD, Solomon AW, Biamonte MA et al (2021) Diagnostics and the neglected tropical diseases roadmap: setting the agenda for 2030. Trans R Soc Trop Med Hyg 115:129–135. https://doi.org/10.1093/trstmh/traa118
Suter E, Oelke ND, Adair CE, Armitage GD (2009) Ten key principles for successful health systems integration. Healthc Q 13(Spec):16–23. https://doi.org/10.12927/hcq.2009.21092
Tagoe ET, Sheikh N, Morton A, Nonvignon J, Sarker AR, Williams L, Megiddo I (2021) COVID-19 vaccination in lower-middle income countries: National Stakeholder views on challenges, barriers, and potential solutions. Front Public Health 9:709127. https://doi.org/10.3389/fpubh.2021.709127
Thuo IW, Ndiaye K, Mookherji S (2020) Factors influencing the sustainability of neglected tropical disease elimination programs: a multi-case study of the Kenya national program for elimination of lymphatic filariasis. Am J Trop Med Hyg 102(5):1090–1093. https://doi.org/10.4269/AJTMH.19-0261
UN. (2015). Global strategy every woman every child.. https://www.everywomaneverychild.org/global-strategy/. Accessed 28 Dec 2022
Velin L, Lartigue J-W, Johnson SA, Zorigtbaatar A, Kanmounye US, Truche P, Joseph MN (2021) Conference equity in global health: a systematic review of factors impacting LMIC representation at global health conferences. BMJ Glob Health 6(1):e003455. https://doi.org/10.1136/bmjgh-2020-003455
Wang H, Liddell CA, Coates MM, Mooney MD, Levitz CE, Schumacher AE et al (2014) Global, regional, and national levels of neonatal, infant, and under-5 mortality during 1990–2013: a systematic analysis for the global burden of disease study 2013. Lancet 384(9947):957–979. https://doi.org/10.1016/S0140-6736(14)60497-9
WHO. (2018). Re: comments on the draft first report of the WHO independent high-level commission on non-communicable diseases.. In بثبثب.. http://apps.who.int/
Woodward A, Sheahan K, Martineau T, Sondorp E (2017) Health systems research in fragile and conflict affected states: a qualitative study of associated challenges. Health Res Policy Syst 15(1):1–12. https://doi.org/10.1186/s12961-017-0204-x
World Health Organisation (WHO) (2016) Neglected tropical diseases. https://doi.org/10.1037/0021-9010.93.1.170
World Health Organization - Regional Office for Europe (2017) Migration and health: key issues. WHO Regional Office for Europe
World Health Organization (WHO) (2017) Together on the road to universal health coverage: a call to action. World Health Organization, pp 1–38. https://apps.who.int/iris/bitstream/handle/10665/258962/WHO-HIS-HGF-17.1-eng.pdf;sequence=1%0Ahttp://www.who.int/universal_health_coverage/road-to-uhc/en/%0Ahttp://apps.who.int/iris/bitstream/10665/258962/1/WHO-HIS-HGF-17.1-eng.pdf?ua=1
Zafar SN, McQueen KAK (2011) Surgery, public health, and Pakistan. World J Surg 35(12):2625–2634. https://doi.org/10.1007/s00268-011-1304-3

3 Priority Diseases, Sustained Response and SDG3.3

Abstract

Priority diseases are those designated by WHO as having potential to cause a public health emergency. As such effective countermeasures against these diseases are crucial both at National and global level. The list of priority diseases is updated regularly by WHO. The diseases are rated critical, high, and medium priority and they range from bacterial, viral and fungal pathogens. The fungal pathogen priority list was compiled with a focus on the invasive diseases, limited diagnostics and treatment, and antifungal resistance. However bacterial priority diseases have been ranked according to their antibiotic resistance. Over 700,000 deaths annually have been attributed to drug resistant bacteria. The ability of viruses to spillover from animals to humans and their ability to cause disease in humans is a public health threat globally. In the past decades emerging and re-emerging diseases from viral pathogens of animal origins have been reported. The lack of treatment and vaccines for most viral priority diseases makes these diseases public health threats globally.

Keywords

Priority disease · Fungi · Bacteria · Virus · Pathogens · Low- and Middle-Income Countries · Crimean Congo haemorrhagic fever · Sustainable development goals (SDG) · Spillover · Sustainable healthcare

3.1 Priority Diseases

The World Health Organization has published a list of priority diseases that is updated on a regular basis. These priority diseases are caused by various pathogens with viral pathogens being implicated in most of the recent outbreaks including the COVID-19 pandemic (Table 3.1). A list of priority bacterial and fungal pathogens has also been published ranging from critical, high to medium priority (Tables 3.2 and 3.3). Publication of these priority diseases is crucial in addressing SDG3d "Strengthen the capacity of all countries, in particular developing countries for early warning, risk reduction and management of national and global health risks". The capacity of LMIC in management of priority diseases varies considerably with financial resources and availability of necessary infrastructure at the core of preparedness and response.

3.1.1 Priority Diseases from Viral Pathogens

One common aspect among the priority diseases caused by viral pathogens is the zoonotic aspect

F. A. Mphande, *Sustainable Health in Low and Middle Income Countries*, Sustainable Development Goals Series, https://doi.org/10.1007/978-981-99-4254-1_3

Table 3.1 List of viral priority diseases

Priority disease	Case fatality	Vaccine
Crimean-Congo haemorrhagic fever (CCHF)	10–40%	No vaccine available for human and or animals
Ebola virus disease	50%	Available (Ervebo, Zabdeno-and-Mvabea)
Marburg virus disease	23–100%	No approved vaccines and or therapies available
Lassa fever	1% and 15% in severe cases	No vaccine available
Middle east respiratory syndrome coronavirus (MERS-CoV) severe acute	Approximately 35%	No vaccine or specific treatment available
Respiratory syndrome (SARS)	11–15%	No vaccine available
Nipah and henipaviral diseases	40–75%	No treatment or vaccine available
Rift Valley fever (RVF)	Not reported, mostly causes mild disease in humans	Vaccine for humans (not licensed) is available for high-risk personnel (laboratory and veterinary personnel). Vaccine for animals is available
Zika	Not reported	There is no treatment and no available vaccines
Disease X (represents a serious international epidemic caused by a pathogen that is currently unknown to cause disease in humans)	Not reported	N/A

of the diseases. The ability of viruses to spillover from animals to humans and their ability to cause disease in humans is a public health threat globally. In the past decades emerging and re-emerging diseases from viral pathogens of animal origins have been reported. Outbreaks from Nipah virus, Ebola haemorrhagic fever, Marburg haemorrhagic fever and the current COVID-19 pandemic are evidence of how spillover can have deadly consequences in humans (Table 3.1). While some diseases such as Ebola had case fatality rates up to 50%, Nipah virus, 40–70% and MERS-CoV up to 30% and CCHFV, 30%, most of these diseases do not have treatment and or vaccines available (Bower et al. 2019; Dixon and Schafer 2014; Gómez Román et al. 2020). The lack of treatment and vaccines for most viral priority diseases makes these diseases public health threats globally.

Table 3.2 List of fungal priority pathogens

Priority group	Pathogen
Critical priority group	*Cryptococcus neoformans*
	Aspergillus fumigatus
	Candida auris
	Candida albicans
High priority group	*Nakaseomyces glabrata*
	Eumycetoma
	Fusarium spp.
	Candida parapsilosis
	Histoplasma spp
	Mucorales
	Candida tropicalis
Medium priority group	*Scedosporium spp.*
	Lomentospora prolificans
	Coccidiodes
	Pichia kudriavzeveii (Candida krusei)
	Cryptococcus gattii
	Talaromyces marneffeii
	Pneumocystis jirovecii
	Paracoccidioides

3.1.2 Priority Diseases of Fungal Origins

Priority diseases caused by fungal pathogens should not be underrated. Apart from viral pathogens, fungi can cause superficial infections, invasive, acute and subacute infections which can

Table 3.3 List of bacteria priority pathogens

Priority category	Pathogen	Antibiotic resistance
Critical	*Acinetobacter baumannii*	Carbapenem-resistant
	Pseudomonas aeruginosa	Carbapenem-resistant
	Enterobacteriaceae	Carbapenem-resistant
	Klebsiella pneumonia	Third generation
	Escherichia coli	Cephalosporin-resistant
	Enterobacter spp.	
	Serratia spp.	
	Proteus spp.	
	Providencia spp.	
	Morganella spp.	
High	*Enterococcus faecium*	Vancomycin-resistant G
	Staphylococcus aureus	Methicillin-resistant vancomycin intermediate and resistant
	Helicobacter pylori	Clarithromycin-resistant
	Campylobacter	Fluoroquinolone-resistant
	Salmonella spp.	Fluoroquinolone-resistant
	Neisseria gonorrhoeae	Third generation cephalosporin-resistant fluoroquinolone-resistant
Medium	*Streptococcus pneumoniae*	Penicillin-non-susceptible
	Streptococcus pneumoniae	Ampicillin-resistant
	Shigella spp.	Fluoroquinolone-resistant

result in severe infections often times with fatal outcomes. Invasive fungal infections are on the increase with immunocompromised individuals at high risk. During the COVID-19 pandemic, an increase in cases of mucormycosis an invasive fungal infection caused by Mucorales (High priority pathogen) was observed and immunosuppression was implicated in the increase in numbers of cases (Sarfraz et al. 2022; Selarka et al. 2021).

The fungal pathogen priority list was compiled with a focus on the invasive diseases, limited diagnostics and treatment, and antifungal resistance. Unlike most viral and bacterial pathogens, fungal pathogens are extremely understudied as such there is limited data on the distribution patterns of these pathogens as well resources and quality data, hence difficult to estimate the disease burden (World Health Organization 2022).

The priority list was compiled according to number of deaths, annual incidence, current global distribution, trends in the last 10 years, inpatient care, complications and sequelae, antifungal resistance, preventability, and access to diagnostic tests.

While strategies such as surveillance are crucial in reducing the knowledge gaps, access to equipment that can assist in better and effective diagnostics is essential. In a paper by Benedict et al. it was estimated that in 2017, $7.2 billion was spent on cost of fungal treatment, including $4.5 billion on hospitalizations and $2.6 billion from outpatient visits (8,993,230). *Candida* infections accounted for 26,735 patients with a total cost $1.4 billion, and *Aspergillus* infections, 14,820 total visits at a total cost $1.2 billion (Benedict et al. 2019). While HIC can afford such costs, this is close to impossible in many LMIC. Diagnostic tests such as MALDI-TOF (matrix-assisted laser desorption/ionization time of flight) mass spectrometry systems, real-time PCR (polymerase chain reaction) are available in most HIC, but LMIC are restricted to basic microscopy and culture, if they a have a microbiology laboratory. This creates a huge barrier in making the right diagnosis and providing the right treatment. Basic microscopy and culture limit the types of fungal organisms that can be detected, as most invasive fungal species are dangerous to be handled in culture especially in a basic microbiology laboratory. Apart from making the right diagnosis, antifungal therapeutic monitoring is another challenge for LMIC due to high costs thus is limited mostly to high-income settings (Benedict et al. 2019; Terrero-Salcedo and Powers-Fletcher 2020; World Health Organization 2022). For the available therapies and treatment for fungal infections, uneven distribution of these treatments has affected treatment of these infections in LMIC. Lack of diagnostic tools, surveillance and microbiological confirmation of pathogens has contributed to misdiagnosis, under reporting, and poor fungal pathogen distribution data respectively. Low- and middle-income countries have (if any) limited susceptibility data hence not able to recognize antifungal resistance patterns and distribution within their territories (Jajoo et al. 2018).

3.1.3 Priority Diseases of Bacterial Origin

Apart from viral and fungal pathogens, bacterial pathogens are another public health threat, not on the aspect of emerging disease like viral infections but more on antimicrobial resistance. Over 700,000 deaths annually have been attributed to drug resistant bacteria. Antibiotic resistant bacteria is spreading worldwide affecting the treatment of many bacterial diseases that were treatable. For example cases of multi drug resistant TB have been reported worldwide further complicating the treatment of the disease (Belkina et al. 2014; Lukoye et al. 2013). A list of priority bacterial pathogens was published by WHO to assist in alerting which pathogens are resistant to which antibiotics (Table 3.3). The pathogens have been given different categories of priority, with a critical priority category given to pathogens resistant to Carbapenem (*Acinetobacter baumannii, Pseudomonas aeruginosa, Enterobacteriaceae*), cephalosporin (*Escherichia coli)* and Third generation (*Klebsiella pneumonia)*. A high priority has been allocated to bacterial pathogens resistant to vancomycin (*Enterococcus faecium)*, Methicillin (*Staphylococcus aureus*), Clarithromycin (*Helicobacter pylori*), flouroquinolone (*Campylobacter, Salmonella spp*.,) and Third generation-cephalosporin (*Neisseria gonorrhoeae*). Medium priority given to penicillin non-susceptible (*Streptococcus pneumoniae*), Flouroquinolone resistant (*Shigella spp.*), and Ampicillin resistant pathogens (*Streptococcus pneumoniae*) (Table 3.3).

3.2 Management of Priority Diseases

Management of priority diseases requires understanding of the various pathogens, their epidemiology as well as molecular, biochemical and immunological data. Different settings and pathogens require approaches that are applicable as well as effective to the setting and the pathogen.

3.2.1 Management of Priority Viral Diseases

While some of the priority diseases are emerging diseases others are re-emerging, with various animal hosts as reservoirs of the disease-causing pathogens. The relationship between the reservoir host and the human host relies in the interaction between humans and the reservoir host and or vectors that carry the disease from the reservoir host to the human host. Human-to-human transmission of disease also occurs resulting in the spread of the disease within the human population. If there is sustained human-to-human transmission, this could result in an outbreak situation which could escalate into an epidemic and worse still into a pandemic.

When thinking of priority diseases, it is important to consider the host, vector, the reservoir, migratory animals, and the environment as these play a role in emergence and re-emergence of infectious disease. In the case of CCHF an infected tick can spread the virus through vertical transmission from larva, nymph, and adult stage.

With over 70% of emerging diseases being zoonotic diseases, jumping from wild animals to humans as well as domestic animals, the threat of emerging and re-emerging diseases is eminent. While the scientific community are working hand in hand to understand and combat these diseases, further cooperation is required between disciplines, governments, and institutions globally. With platforms such as PROMED, information is made available at what is happening around the world regarding infectious disease outbreaks in plants, animals, and humans. Although this information is readily available, not everyone is able to access and understand it. How can this information be more available to the public? How can the public engage in a positive way in adopting early warning systems, outbreak management and response? How can people access the right information in this time of conspiracy theories? How can LMIC as well as communities that are at risk of emerging diseases benefit from early warning systems in their outbreak preparedness, management, and response?

Countries with limited resources have struggled to collect data effectively as such it is a challenge to keep track of various priority pathogen risks within and around the territories. Effective and sustainable surveillance systems are thus crucial to achieving better preparedness. Availability of surveillance data is vital in the development of early warning systems; and lack of it considerably affects preparedness. SDG3d is thus crucial for LMIC which carry a huge burden of emerging and re-emerging diseases, NTD as well as some of the listed priority diseases.

In the past decade there have been outbreaks from several priority pathogens across all WHO regions. Ebola, CCHF, Monkeypox MERS, SARS and the current Marburg outbreaks have affected populations worldwide, but LMIC have been affected most. With limited data, LMIC have struggled with preparedness and response, most often because the response is reactive than proactive. With limited resources, priority is given to ongoing challenges than anticipated ones (Engels and Zhou 2020; Feldmann et al. 2018; Haq et al. 2013; Khan and Qazi 2013; Perry et al. 2014). It is thus timely to encourage LMIC to focus on emergency preparedness, and early warning systems for the various outbreak/epidemic prone diseases. To achieve early warning, there is need for sufficient data that could be used to observe trends and patterns of disease occurrence within the communities. By understanding these trends and linkages risk and protective factors could be identified and early warning systems can be put in place. While all this is doable, the challenge in LMIC still remains availability of data. It is therefore essential that sustainable surveillance systems are built within LMIC as there are key to developing early warning systems. Improving digital technology and access to internet in LMIC could assist in data collection, analysis, management, and dissemination. Early warning systems could assist in risk reduction while improved digital technology could assist in building sustainable early warning systems. While internet access is still limited, access to mobile phones is far reaching in both urban and rural populations in LMIC. Thus, tapping into existing networks that are well established within communities in LMIC could be beneficial for a robust and resilient early warning system against disease outbreaks and other health emergencies. Community health workers (CHW) are a key link between healthcare professionals and the communities in LMIC. Despite working often on a voluntary basis, CHW reach places where other healthcare professionals barely reach (Abrahams-Gessel et al. 2015; Awah et al. 2018; Haroun et al. 2022; Nuwematsiko et al. 2022; Perry et al. 2014; Stewart et al. 2018). With proper training and the right equipment, CHW could be critical in collecting real time surveillance data from the various communities. Thus CHW, could be one network to tap into in LMIC for surveillance.

Viral pathogens are capable of replicating in host animals and crossing over into humans and mutate at high rates which contributes to challenges in managing these diseases. Spillover events from viral pathogens have resulted in disease outbreaks, epidemics and pandemics in the last century.

Understanding the sources and drivers of zoonotic diseases within the population is crucial in disease prevention, management and control. The drivers of zoonotic diseases include the environment/geographical setting, the animal reservoir hosts (domesticated and or wild), and the human host. Many studies have been conducted in different disciplines including environmental health, animal health and human health to understand the mechanisms that drive diseases. Most of these studies were done separately, but most recent, multi-disciplinary and multi sectoral approaches have paved way to better understanding of the drivers of disease. A One Health Approach was recommended in 2007 as a pandemic preparedness response at the International Ministerial Conference on Avian and Pandemic Influenza in New Delhi. Following the recommendation, FAO, WOAH, WHO, UNICEF, the World Bank, and the United Nations System Influenza Coordination (UNSIC) developed a strategic framework called "Contributing to One World, One Health™-A Strategic Framework for Reducing Risks of Infectious Diseases at the Animal-Human-Ecosystems Interface." The strategy applied the One Health concept to emerging infectious diseases at the animal-human-ecosystem interface following lessons learned from the highly pathogenic H5N1 avian influenza outbreak response in early 2000 (https://www.cdc.gov/onehealth/basics/history/index.html). Currently the One health approach is recognized as an important aspect in addressing diseases that cross the animal-human interface with many countries taking on this approach. The multisectoral and multidisciplinary approach is important in building mechanisms for coordination, communication, and collaboration both at national and international level to address global health threats at the human-animal environment interface. One health approach plays an important role in the Sustainable Development Goals 2030 Agenda (Clark et al. 2021; Elphick-Pooley and Engels 2020; Kuruvilla et al. 2018).

3.2.2 Management of Priority Fungal Diseases

Management fungal pathogens is an ongoing discussion has led to the development of guidelines to assist in combating fungal infections globally (Thompson et al. 2021). Management of fungal infection includes infection prevention and control, treatment, diagnosis, combating anti-fungal resistance, and access to better diagnostic and treatment tools. Novel therapies, novel and improved diagnostic and culture methods. All these require availability of resources to acquire the required tools. Direct healthcare costs for treatment of fungal infections both superficial and invasive infections are beyond reach for most LMIC. Novel therapeutic strategies require molecular, biochemical and immunological techniques which are more sensitive, rapid but costly hence out of reach for most LMIC (Benedict et al. 2019; Terrero-Salcedo and Powers-Fletcher 2020). Currently, proven diagnosis requires microbiological and or histopathologic tests aimed at providing clarity and uniformity globally but not all countries are able to access such techniques (De Pauw et al. 2008). Conventional methods which include microbiology and culture have been widely used but have wide limitations including low sensitivity, slow turn around (long periods of culture) and are weak in diagnosis of invasive infections (Terrero-Salcedo and Powers-Fletcher 2020). Non-culture methods such as detection of fungal antigens, antibodies as well as nucleic acids are more specific, rapid, and crucial in clinical decision making. Despite these qualities, these methods still have limitations including limited availability and may not be appropriately used in most in LMIC (Nguyen et al. 2012).

Improving diagnostic methods including culture as well as introducing novel detection methods that can be applied in low-income setting would assist in detecting fungal pathogens, treatment, surveillance, and reporting of fungal infections. With regular and accurate reporting from surveillance at national level, there are greater chances of understanding the distribution of fungal pathogens and associated diseases. Culture method remains the gold standard for diagnosis of invasive fungal infections, but lack of sensitivity and lengthy incubations remain a challenge.

Inclusion of selective media in routine microbiological testing could improve detection of fungal pathogens which often go undetected. This would enhance availability of fungal culture surveillance data especially in LMIC where limited data is available (Hong et al. 2017).

Despite the limitations in availability of fungal detection methods, combining various fungal detection methods could assist in detecting these pathogens in limited resource settings. Identifying a combination of detection methods that can be utilized together and complement each other while enhancing the detection rate of the pathogens could prove valuable for LMIC (Table 3.4).

Quote 3.1

Case Study-Cryptococcus neoformans (Loyse et al. 2019)

Cryptococcus neoformans results in 223,100 cases of cryptococcal meningitis per year and approximately 70% mortality in LMIC. In sub–Saharan Africa alone, where there is a high number of people living with HIV around 135,900 deaths are recorded annually with no sign of decrease of the disease.

Cryptococcal lateral antigen flow is a sensitive assay that has been used to detect the disease in CSF and Blood. The assay is sensitive and specific and is crucial to the diagnosis of the disease. While the assay is widely used in high income countries, most LMIC where the disease is rampant have no access to this crucial diagnostic tool. Apart from diagnostic tools, LMIC are also lacking in flucytosine an essential drug for the treatment of the disease. Flucytosine is either unavailable in LMIC or in places where it is available within the region, it is dangerous to use due to limited availability of intensive laboratory monitoring and management. Flucytosine and Amphotericin B are the gold standard for treatment of cryptococcal meningitis but are lacking in LMIC. Several limitations have been outlined indicating why these essential drugs are not as readily available in LMIC.

Lack of markets (market failure) in LMIC

Flucytosine is not listed on the essential medicines list in most LMIC

High cost

Shortage of manufacturer's interested in developing a generic drug and for those manufacturing the generic drug, they are not distributing the drug in countries that need it most.

Flucytosine is not registered in any African country.

In South America flucytosine is registered in Brazil and Argentina but only available in Colombia and French Guiana.

This is one example of how LMIC are left behind in the fight against priority diseases (Quote 3.1). *Cryptococcus neoformans* the pathogen responsible for cryptococcal meningitis is listed a critical priority pathogen but places where disease is rampant have no access to effective treatment.

3.2.2.1 Strategies for Management of Fungal Pathogens

Fungal pathogens especially priority fungal pathogens should be of great concern globally. The ubiquitous nature of these microorganisms, their ability to invade through various pathways (through the skin, inhalation, formite) and their ability to propagate far and survive longer (in the form of spores) as well as the limited diagnostic and therapeutic tools are gaps that have to be addressed now than later.

Surveillance of fungal pathogens at national, regional, and global level as well as reporting of these pathogens is essential. Improving diagnostic capacity especially in LMIC is important because of the many challenges associated with access to diagnostic tools, which could imply that many fungal infections go undetected and unreported. This skews the prevalence of fungal infections and contributes to the limited attention given to these infections. Limited attention not only refers to reporting but also the amount of research and research funding attributed to fungal pathogens and diseases. Management of fungal infections requires knowledge of the nature of the pathogen, the epidemiology of the pathogen and associated infection as well as the treatment, host pathogen interactions including host susceptibility. Most fungal infections do not have vaccines as such depend on treatment of the resulting

Table 3.4 Detection methods for fungal pathogens and their availability and usage in LMIC

Method	Advantages	Limitations	Availability/ Usage in LMIC
Microbiological methods (culture and microscopy)	Good for phenotypic identification of various fungal species	Depend on the knowledge and expertise of the clinical microbiologist	Limited
		Time consuming	
		Not all clinically relevant moulds can be identified phenotypically	
Molecular and immunological techniques e.g. sequencing, proteomic and serological methods	Accurate in identification and differentiating fungal species	Limited access knowledge and skills of experts high costs	Limited
DNA sequencing	Fast	The cost of testing is high (for LMIC)	Limited
	High accuracy and specificity	Lack of commercially available technologies for routine use in clinical laboratories	Limited
		Challenges in databases where these sequences can be curated and limited access for LMIC	Limited
MALDI-TOF MS	Effective in identifying yeast species	The cost of testing is high (for LMIC)	Limited
		Lack of commercially available technologies for routine use in clinical laboratories	Limited
Fungal antigen detection 1,3, β-D-glucan (BDG)	A pan-fungal antigen found in several antigens including *Candida* species, *pneumocystis jirovecii, aspergillus spp., fusarium spp., acremonium spp,* except *Mucorales, Blastomyces dermatitidis,* and *cryptococcus spp.*	High cost	Limited
	FDA approved for use in serum specimens	Limited availability in LMIC	

infection. Treatment of fungal infection could take a few days to months or years depending on the extent of the infection. In severe cases debridement and amputation is required resulting in morbidity and in some cases mortality. Systemic fungal infections can cause serious disease and are often fatal, hence early diagnosis and availability of an effective treatment makes a huge difference between survival and death. Drug resistance and limited access to effective antifungal drugs have plagued LMIC and have negatively affected the quality of life for those living with fungal infections. Limited access to therapeutics can be attributed to several factors including, cost of drugs, licensing as well as lack of sustainability in acquiring therapies and managing routine fungal infections. Sustainable investment research R&D of antifungal medicines and fungal infections in general could generate relevant therapeutics and data that could be used to make informed decisions. The data generated could assist in meeting the need for public health intervention and to highlight the importance of fungal infections. Including through incorporating fungal diseases and priority pathogens in clinical (medical) and priority public health training and curricular at all levels of training is essential in emphasizing the importance of these pathogens. Drug resistance has been associated with inappropriate use and handling of medicines, as such collaboration across sectors is required to address the impact of antifungal use on resistance across one health spectrum. Here the one health spectrum looks at multisectoral approaches that involve various stakeholders from health, environmental, agricultural, climate change as well as communities around the world. Considerations of regional and national contexts need to be addressed while implementing these programmes.

3.2.3 Crimean–Congo Hemorrhagic Fever Virus (CCHFV)

With a fatality rate of approximately 40%, Crimean-Congo haemorrhagic fever (CCHF) virus is a cause of public health concern. Capable of causing severe viral haemorrhagic fever outbreaks, the disease is transmitted to people through ticks and livestock animals. Contact with bodily fluids such as blood and other secretions from an infected person can result to human-to-human transmission. By 2018 CCHF had been reported in endemic countries in Africa, the Balkans, the Middle East and Asia, in countries south of the 50th parallel north including India and Spain. There is no vaccine available for either people or animals.

Crimean–Congo hemorrhagic fever virus is a Tick-borne disease that affects both animals and humans. Characterized by heamorrhagic fever both in animals and humans, CCHFV can cause severe and often fatal disease with outbreak fatality rates ranging from 3 to 80% (Kasi et al. 2020; Leblebicioglu et al. 2016; Mostafavi et al. 2017; Hotez et al. 2012; Shahbazi et al. 2019).

The disease was classified as A list priority disease by WHO in h due to its capacity of human-to-human transmission, high mortality rate, and a lack of effective therapeutics and or vaccines for human or animal use (World Health Organization 2020). The disease has been reported in over 30 countries in Africa, Asia, Southeast Europe, and the Middle East (Fig. 3.1) (Table 3.1).

Ticks from the Hyalomma species are the known reservoir and vector of CCHF. The ticks transmit the disease both to humans and animals through tick bites, contact with body fluids of domestic animals or patients with Crimean–Congo hemorrhagic fever (CCHF) and crushing of engorged ticks (Papa et al. 2015; Shahbazi et al. 2019).

The disease can be in circulation between tick and vertebrate host without being noticed by the public creating a greater chance of potential spillover, as such prevention of spillover should be a greater priority. So how can this spillover be prevented?

There are several ways that spillover can be prevented, this includes transmission control, surveillance, as well as various multisectoral and multidisciplinary approaches involving the environmental, animal and human health (Fillâtre et al. 2019; Sorvillo et al. 2020).

3.2.3.1 Transmission of CCHFV

Transmission of CCHFV spans three areas, these are environmental health, agriculture/land use and climate change. These three areas play a role in tick-host interaction, in this case the animals whether domesticated and or wild animals (Fillâtre et al. 2019).

Crimean Congo Haemorrhagic fever is a zoonotic disease as such apart from animal health, environmental health, human health are also affected. As humans interact with animals, they are exposed to virus infested animal blood, tick bites, and crushing of ticks (Sorvillo et al. 2020). Human-to-human transmission at household level or through hospital acquired infections (nosocomial infections) especially in settings with limited access to PPE or poorly fitted PPE (Fillâtre et al. 2019; Sorvillo et al. 2020).

There have been reports of transmission of the CCHFV during mating or co-feeding as well as transovarial transmission, in adult *H. truncatum* ticks. These mechanisms represent additional mechanisms that may contribute to the mainte-

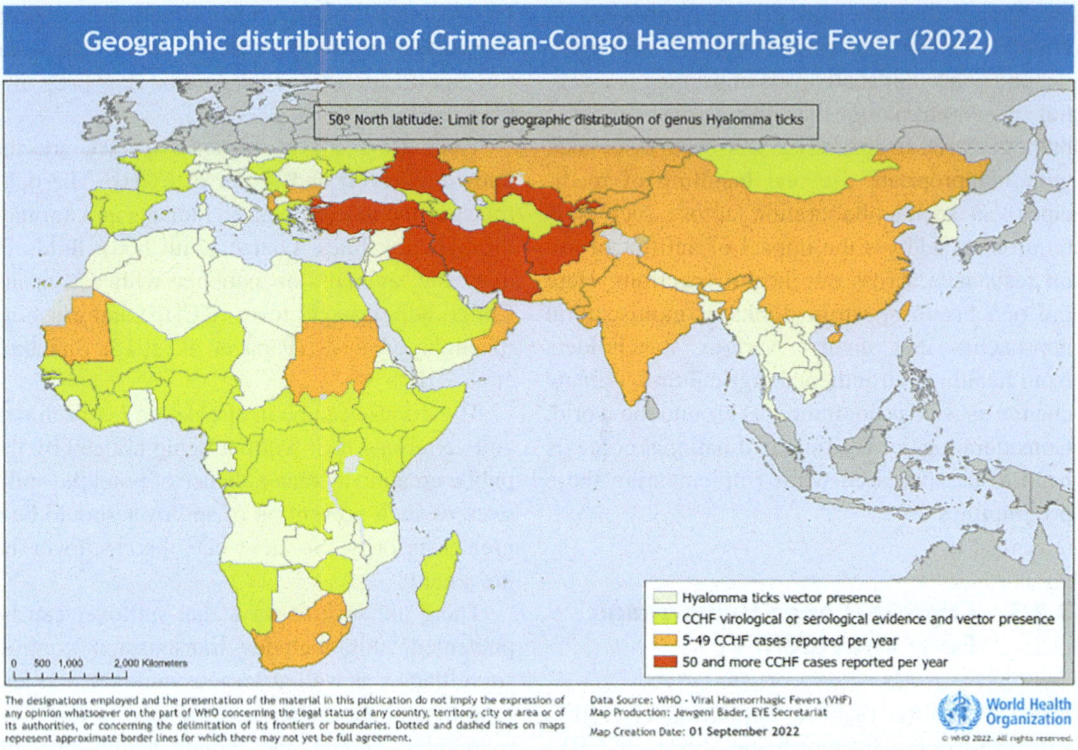

Fig. 3.1 Geographic distribution of Crimean-Congo haemorrhagic fever (2022). Source: https://www.who.int/health-topics/crimean-congo-haemorrhagic-fever/#tab=tab_1

nance of transmission in nature (Gonzalez et al. 1992).

3.3 Disease Prevention Strategies

Since CCHFV not only involves humans but also animals as well as the surrounding environment, there is need for involving both animal and human and environmental experts in designing prevention strategies (Al Awaidy and Al Hashami 2020; Sorvillo et al. 2020; Zakham et al. 2019).

These strategies include, surveillance, risk assessment, and risk reduction strategies focused on humans, animals and ticks. There is need to identify and highlight gaps in knowledge to attain more sustainable disease prevention and control for CCHFV both in animals and humans (Sorvillo et al. 2020).

Multisectoral One Health approach to Disease Control

According to WHO, One Health is a collaborative approach with multidisciplinary, and multisectoral involvement that can address urgent, ongoing, or potential health threats at the human-animal-environment interface at subnational, national, global, and regional levels. Various sectors can be involved in this approach (World Health Organization (WHO) 2019).

The Food and Agriculture Organization of the United Nations (FAO), the World Organization for Animal Health (oie), and WHO published a “A Tripartite Guide to Addressing Zoonotic Diseases in Countries” in 2019. The Framework is focused on One Health Approach at it includes (1) strategic planning and emergency preparedness; (2) surveillance for zoonotic diseases and information sharing; (3) coordinated investigation and response; (4) joint risk assessment for zoonotic disease (World Health Organization (WHO) 2019).

One Health Approach is an important component in both preparedness and response to zoonotic disease threats (Table 3.5). Effective preparedness allows for prompt and timely response to various emergencies be it at national and global level. Challenges in accountability and lack of effective response are some of the drawbacks that have been observed in the recent outbreaks especially with the COVID-19 pandemic. Sharing of information was a challenge and at times led to stigma with costly consequences for the affected countries and communities. Such experiences could negatively affect response. Guidelines, policies, and regulations are crucial in implementing response not only at national but also global level. Despite the differences in income, availability of tools that could aid in effective response and inequalities, most countries worked hard to respond effectively to the COVID-19 pandemic with variable outcomes. Countries were only as strong as their finances and health systems would carry them to mount an effective response for their respective populations. High income countries were quick to mobilize resources to support their populations socially, economically and health. On the other hand LMIC struggled to mobilize resources and to mount an effective response (Burkle 2020; Caballero-Anthony 2021; Chua et al. 2020; Contini et al. 2020; Jowell and Barry 2020; McClarty et al. 2022; Zinsstag et al. 2020).

Table 3.5 Zoonotic disease threats, One Health, Preparedness and Response

Preparedness	Response
Timeliness and effectiveness of response; *early and or timely preparations ensure a prompt and effective response*	Response to zoonotic disease events and emergencies
Availability of information in all sectors; *this ensures that sectors are able to identify gaps and are prepared on time with effective tools to tackle the various challenges and needs*	Decisions are based on accurate and shared assessments of the situation
Accountability to each other and to decision makers ensures action by all sectors; *accountability is essential as it allows for various sectors to take responsibility in their respective areas of preparedness and the quality and effectiveness of the response*	Effective collaboration
Regulations, policies, and guidelines are realistic, acceptable, and implementable by all sectors; *all sectors should review the various policies regulations and guidelines in order to pave way for an effective response. This includes implementation and adherence to the response measures*	All sectors understand their specific roles and responsibilities in the collaboration
Technical, human, and financial resources are effectively used and equitably shared; *this should be considered both at national and global perspective with consideration of vulnerable populations and communities*	Gaps in infrastructure, capacity and information are identified and filled; *huge gaps still exist between high income LMIC as well as at national level, between urban and rural populations including vulnerable and neglected populations*
Advocacy for funds, policies, and programmes is more effective	

3.3.1 Considerations for Sustainability of Implementation of a Multisectoral, One Health Approach

While the One Health Approach framework is readily available, and most countries are able to effectively implement it, sustainability remains an issue. It has been observed that during the emergency, the approach will be implemented often with much success, but the moment the threat has been neutralized the approach is abandoned and becomes a story of the past. So how do countries ensure sustained preparedness and response to zoonotic disease threats? Political will, governance, and availability of resources are some of the challenges that directly or indirectly affect preparedness and response to disease outbreaks including zoonotic diseases (Akseer et al. 2020; Assefa and Gilks 2020; Kelly-Hope et al. n.d.; Gayer et al. 2007; World Health Organization (WHO) 2019).

Political will: At national level, political will and engagement from national leaders is very crucial for sustainable One Health Approach. Political will allows for availability of resources both financial and human capital to sustain the different zoonotic disease prevention and control programmes. Involvement of relevant sectors both government and non-governmental organizations is of ultimate importance (Kelly et al. 2017; World Health Organization (WHO) 2019).

Resources: For a programme to be sustainable there is need for resources, including financial resources, human resources and infrastructure at national level. Distribution of these resources will determine the level of preparedness and response to threats for the country as a whole but also in induvial communities within the country. In places where there is political unrest, corruption and widening social divide, equitable distribution of resources is a challenge often with undesirable outcomes. This may result for example the disease affecting populations according to the social divide with the poor and vulnerable carrying most of the disease burden. At global level, equitable distributions of resources is yet to be attained. The distribution of treatment and disease prevention resources such as vaccines and medicines for COVID-19 and Monkeypox are still skewed with access in high income countries while LMIC are surviving with limited and at time no access (Burkle 2020; Kabuga and El Zowalaty 2019).

Context/Setting: Consideration of the social and geographical environment where One Health Approach will be implemented is crucial. This not only guides the policy makers on what is needed and required for effective implementation of the program, but also how these interventions will be implemented in the target population and geographical setting. Planning activities within the existing infrastructure both at national and community level could assist in successful implementation of the programmes. Varying infrastructure and resource capabilities greatly affect how each setting can respond to disease and emergencies (Deml and Githaiga 2022; Jayatilleke 2020; McClarty et al. 2022; Vaughan et al. 2018).

Common goals: As a collaborative approach, a shared common goal between various stakeholders and sectors could ensure adherence and support which could be essential in commitment to the programme and greater chances of sustainability. Designing strategies that incorporate shared needs, benefits health priorities between stakeholders plays an important role in sustainability of the common goal (Dungu et al. 2018; Gambhir et al. 2015; Jowell and Barry 2020; Meltzer et al. 2016; Petersen et al. 2019; Pigott et al. 2017).

Strong governance: For any intervention program be it economic, social or health related to be successful, governance is key. Poor governance hinders not only implementation but also adherence and sustainability of the program. For health systems, strong national governance structures are thus required to ensure that policies are followed, as well as ensuring that required frameworks are available to guide the population. Governance is also important in ensuring that there is compliance to the existing standards at national, regional and global level and that these standards are adhered to (Gilson et al. 2017).

Routine coordination and communication: Multi sectoral collaboration involves working with people from different disciplines and programs, as such effective communication and coordination is crucial. While effective coordination assists with planning and implementation of projects, effective communication ensures that information is passed on effectively without misunderstandings. Establishing routine communication between stakeholders ensures that all teams are up to date with the various strategies, activities and plans for the programme (World Health Organization (WHO) 2019).

Recognizing successes: highlighting the success in the various sectors and the programme as a whole serves as a motivation to the members. While recognizing successes, gaps and challenges should be identified and strategies set out for appropriate intervention.

3.4 Sustainable Development Goals and Priority Diseases

Sustained response does not only refer to response to disease and outbreak but also to other social and economic factors affecting the population (Fig. 3.2). In most LMIC, the issues related to disease are often intertwined with social, political, and economic factors that directly and or indirectly affect health outcomes of populations. Understanding the connection between social, economic, political, health and environmental factors and their role in health outcomes is of great importance (Fig. 3.2).

It is interesting to note that not only SDG3.3 should be considered when addressing priority diseases, almost all SDGs directly and or indirectly linked to priority diseases. A summary of SDG1, SDG2, SDG4, SDG5, SDG6, SDG9, SDG10, SDG11, SDG12, SDG13, SDG15, SDG16 and SDG17 and their link to SDG3.3 is highlighted.

SDG1 End poverty in all its forms everywhere. Poverty has always been associated with disease globally. Populations in LMIC carry most of the disease burden associated with poverty including malaria, TB, HIV and NTD. Most of the emerging and re-emerging diseases have been reported in LMIC with populations with limited resources being the most affected (Hotez et al. 2015; Mphande 2016a, 2016b; Oramasionwu et al. 2011; Seimenis 2012; Song et al. 2014; Tomley and Shirley 2009; Undurraga et al. 2017; Wagstaff et al. 2003; World Health Organisation(WHO) 2016; Zere et al. 2007; Zinsstag et al. 2020). The COVID-19 pandemic has further plunged many into poverty with over 4 years of progress being erased (Buheji et al. 2020; Pereira and Oliveira n.d.; Yonzan et al. 2022). Priority diseases such as Nipah, Ebola, Lassa fever often affect rural populations most of whom have limited resources and challenges in access to health facilities (Borremans et al. 2019; Bourgarel et al. 2010; Hallam et al. 2018; M., G., D., L., P., F.,, and M.A., C. 2007).

SDG2 End hunger, achieve food security and improved nutrition and promote sustainable agriculture. Lack of adequate food and or poor nutrition has a great impact on human health. Poor nutrition results in stunted growth, reduced immunity. Hunger has pushed populations to migrate into uncharted territories such as forests in search for fertile lands. This brings these populations into close proximity with wild animals and other reservoirs of pathogens that could potentially spillover into the human population. Over 70% of emerging diseases have been associated with zoonotic origins (Bonwitt et al. 2018; Borremans et al. 2019; Goldspink et al. 2015; Griffin 2010). Diseases such as CCHF have been associated with handling of infected wild and domesticated animals both live as well as dead as well as forest areas (Mertens et al. 2013; Spengler et al. 2016; Wasfi et al. 2016). The risk of CCHF is very high among affected populations as wild and domesticated animals are prone to spreading the disease further globally. With no treatment and very few vaccines for animals and risk populations, sustainable agriculture should consider reducing such risks (Mertens et al. 2013; Spengler et al. 2016; Wasfi et al. 2016).

SDG4 Ensure inclusive and equitable quality education and promote lifelong learning opportunities for all. The Covid-19 pandemic affected school going populations from primary school up to university. The COVID-19 pandemic resulted millions of children missing school and some with no possibility of returning to school (https://sdgs.un.org/goals). Education is crucial in understanding health and wellbeing. Thus, poor education could have negative consequences on the health outcomes of the affected populations (Beattie et al. 2015; Chalem et al. 2007; Stoner et al. 2017; Tice 2013). During the COVID-19 pandemic, it is estimated that 147 million children missed school with approximately 24 million learners who will be unable to return to school (https://sdgs.un.org/goals/goal4). Increase

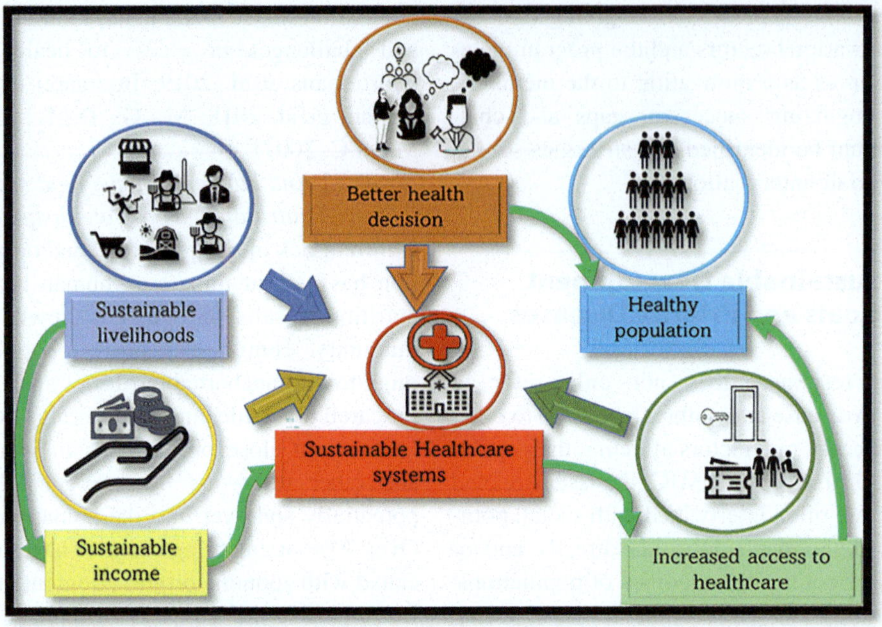

Fig. 3.2 Sustainable Healthcare Systems; sustainable income, sustainable livelihoods, and better health decisions could contribute to a sustainable healthcare system and a healthy population with better access to healthcare

in new HIV cases was reported in women and girls in 2021 (Joint United Nations Programme on HIV/AIDS 2022). Thus, outbreaks associated with any of the priority diseases could have effects on SDG4.

SDG5 Achieve gender equality and empower all women and girls. Child marriages are one of the challenges observed in limited resource populations. This exposes young girls to HIV, sexually transmitted infections and further plunges this young population into poverty. The 2022 UNAIDS report revealed that 63% of the new HIV infections were among women and girls (Joint United Nations Programme on HIV/AIDS 2022). It is tempting to associate the challenges brought about by the COVID-19 pandemic and whether these could have contributed to these new cases (Bundervoet et al. 2021; Iglesias Martínez et al. 2022; Mukherjee 2020). Women have suffered stigma after recovering from diseases such as Ebola negatively affecting their mental health and wellbeing ((CDC), Organization, and Health 1998; Accorsi et al. 2005; Chauhan et al. 2020; Ellington et al. 2017; Jamieson et al. 2014). Similarly, Zika infections had negative pregnancy outcomes with direct consequences on health and wellbeing including social stigma and mental health (Bell et al. 2016; Mead et al. 2018; Shapiro-Mendoza et al. 2017). Priority diseases can thus be linked to SDG5 and failure of SDG5 contributes to challenges on attaining SDG3.3.

SDG6 Ensure availability and sustainable management of water and sanitation for all. Access to clean water is essential for health and wellbeing of populations globally. Lack of clean water is a source of various preventable diarrhoeal diseases including cholera. Cholera outbreaks are endemic in LMIC often with fatal outcomes (Adam et al. 2017; Cauchemez et al. 2013; Farmer 2013; Leclerc et al. 2002; MacPherson et al. 2009; Trolle et al. 2023). With over 1 billion people in LMIC lacking safe clean water, this population is at risk of waterborne diseases including cholera. As such if SDG3.3 is to be attained, SDG6 must be successful and sustainable in LMIC.

SDG9 Build resilient infrastructure, promote inclusive and sustainable industrialization and foster innovation. Poor infrastructure has been linked to challenges in attaining sustainable healthcare in LMIC. For example, infectious disease prevention and control structures such as surveillance systems as well as data collection are in LMIC require strengthening (Alhaji et al. 2018; Gray and Kayali 2009; Sayed et al. 2019). With poor surveillance and lack of data on various diseases including NCD, it is challenging to make informed decisions that could help in management and control of diseases (Aceng et al. 2020; Gell 2015; Jourdain et al. 2019; Marks et al. 2014). Limited resources including health technology and strong health systems have negatively affected health outcomes in populations living in LMIC. Outbreaks from priority diseases could further weaken the already fragile health systems in LMIC (Davey and Burridge 2009; Kollie et al. 2020; Merianos 2007). Investing in research and development as well as innovations could boost capabilities of LMIC to attain SDG3.3.

SDG10 Reduce inequality within and among countries. The COVID-19 pandemic has revealed the extent of health gaps and inequalities globally (Khankeh et al. 2021; Sardar et al. 2020). From the distribution of vaccines and therapies to availability of infrastructure, the gap between HIC and LMIC couldn't be greater(Khankeh et al. 2021; Sbarra et al. 2021). The widening of the gap between these countries needs to be addressed if SDG10 is to be attained. Addressing SDG10 could help reduce the gaps and assist in combating priority diseases globally. Discrimination and migration on various grounds including social status, race, skin colour, religion, political affiliation, disease, as well as geographical has affected treatment, management and control of various diseases (Boulware et al. 2003; Dionne and Turkmen n.d.; George et al. 2020; Knight et al. 2020; Mukherjee 2020; Wang et al. 2021). While refugees and asylum seekers struggle to get treatment from some countries due to their immigration status, others are discriminated against on the grounds of lack of adequate finances to meet their medical needs as well as lack of infrastructure to meet their health needs (Abubakar et al. 2022; Devakumar et al. 2022a, 2022b; Reny and Barreto 2022; Shannon et al. 2022). Migration has allowed for transfer of diseases from endemic to non-endemic countries as well as increased chances of spread of outbreak and pandemic prone diseases and drug resistant pathogens (Das et al. 2006; Kamper-Jørgensen et al. 2012; van Roon et al. 2019). Deep seated inequalities are a major risk factor in attaining SDG3.3.

SDG11 Make cities and human settlements inclusive, safe, resilient, and sustainable. Disparities within countries were magnified during the pandemic (Akseer et al. 2020; Cowling et al. 2013; Hartley 2004; Terefe et al. 2017). While vaccines and healthcare services were readily available in urban areas, rural populations struggled to find these resources (Akseer et al. 2020; Maman et al. 2018; Mutembo et al. 2019). Within the urban populations, people living in slums and other overpopulated settlements struggled with implementation of COVID-19 interventions such as social distancing as well as sanitary housing (Aberese-Ako Id et al. 2022; Akseer et al. 2020; Brotherhood et al. 2022; McClarty et al. 2022; Nuwematsiko et al. 2022). While gender based violence was also exacerbated with the lockdown (Das et al. 2018; Golechha 2020; Mubarak et al. 2016; Terefe et al. 2017). Disparities between urban and rural settlements as well as rich and populations living in poverty negatively contribute to health outcomes. Slums are some of the most populated settlements in urban areas of big cities and they are also the most vulnerable settlements in transmission of communicable disease. The priority diseases outlined in this chapter are highly transmissible, outbreak prone, and often difficult to treat due to lack of therapies. Thus, improvement of such settlements could assist in better health outcomes.

SDG12 Ensure sustainable consumption and production patterns, SDG13 Take urgent action to combat climate change and its impacts, SDG15 Protect, restore, and promote sustainable use of terrestrial ecosystems, sustainably manage forests, combat desertification, and halt and reverse land degradation and halt biodiversity loss. As global populations increase, the amount of food required to feed the growing population has also increased. Environmental factors due to climate change have resulted into droughts which have contributed to food scarcity and famine in some of the most vulnerable populations of the world. Famine has contributed to malnutrition and poor health in affected populations resulting to vulnerability to various diseases including infectious diseases. Displaced populations due to political instability, war and natural disasters often lose their livelihoods and have difficulties in accessing their healthcare needs. The loss of livelihoods affects their ability to have sustainable income, as such are unable to meet their nutritional as well as healthcare needs resulting into poor health. Increase in global population has resulted in large-scale commercial and subsistence farming which involves clearing of indigenous forests to create farms (Farag et al. 2018; Hyder et al. 2023; Kasi et al. 2020). This practice provides an opportunity of interaction between humans and plant and animal populations thus higher likelihood of exposure to new pathogens. The exposure could likely result in spillover events with possibility of pathogens invading and causing disease in humans. Change in lifestyle was reported to encourage spillover events for MERS Co-V from camels to humans (Farag et al. 2018). Similarly the COVID-19 pandemic led to increase of backyard farming e.g. in Cambodia, thus increasing the chance of spillover events between domesticated animals and humans (Hyder et al. 2023).

Apart from potential spillover, the clearing of land for has resulted in loss of biodiversity thereby influencing climate change. The loss of biodiversity has also resulted in migration of wild animals into other territories carrying with them potential pathogens. The emergence of ticks for CCHFV in territories where they were not available before is one example (Mertens et al. 2013; Pascall et al. 2020).

SDG16 Promote peaceful and inclusive societies for sustainable development, provide access to justice for all and build effective, accountable and inclusive institutions at all levels. Sustained peace is a crucial aspect in attaining good health and prosperity globally. Sustained peace is essential if SDG are to be attained. Conflict has resulted in displacement of population within countries as well as across borders. If "health for All" is to be attained, peace has to be achieved in conflict zones. Conflicts disrupt health systems, vaccination campaigns and programs as well as routine healthcare visits for various health needs (Aro and Kantele 2018; Bisika 2010; Burkle Jr. 2017; Haq et al. 2013; Woodward et al. 2017). Displaced populations often lack access to primary healthcare and are at risk of exposure to various disease pathogens, thus are vulnerable to various diseases including priority diseases (Bisika 2010; Burkle Jr. 2017; Woodward et al. 2017). Conflict have contributed to unintended migration of populations across borders resulting in refugees. While some countries are able to provide healthcare to refugee populations, others do not, hence lack of and or limited access to healthcare puts these populations at risk.

SDG17 Strengthen the means of implementation and revitalize the Global Partnership for Sustainable Development. The response to the COVID-19 pandemic has seen strengthening of global partnerships in certain ways but also failure of such partnerships. With the current skewed vaccine coverage, availability of resources and cooperation; trust in global partnerships is hanging by a thread. Strengthening global partnerships would not only assist in attaining SDG3.3 but also combat future outbreaks and or pandemics from priority diseases as well as emerging and re-emerging pathogens. At the national level, countries with sustainable health systems can manage and maintain the healthcare needs for their populations, while those with weak healthcare systems are often overwhelmed by surges such as disease outbreaks and pandemics. Dependence of countries with weaker health systems on donations to manage their healthcare systems, food supply and other financial assistance, have left these nations even more vulnerable as their decision making is often at the mercy of the donor countries' requirements.

3.4.1 Sustainable Healthcare Systems

Sustainable health systems in LMIC is one aspect that countries have to strive if they are to attain universal health coverage. Poverty, poor nutrition, gender inequalities, water and sanitation, lack of resilient infrastructure, global inequalities and poor infrastructure are some of the challenges affecting progress of SDG in LMIC. While working on sustainable health systems, it is crucial to focus on the detailed challenges affecting the systems. By identifying the gaps, disparities and challenges, LMIC could design implementation programmes fitting with the need within their societies instead of one size fits all solutions.

Financial capabilities have limited progress of LMIC and their health systems. While HIC have strong financial capabilities which allows for better and sustained health services for their populations, LMIC are struggling. Most LMIC lack the financial strength that not only can sustain the healthcare system but also ensure a healthy population. According to the concept of Universal health coverage (UHC) by WHO, which translates that all people have access to total quality health services that they need at the time which the services are needed and in their respective setting (where they are needed) (https://www.who.int/health-topics/universal-health-coverage#tab=tab_1). Universal health coverage focuses on a people-centred primary healthcare approach from health promotion to prevention, treatment, rehabilitation and palliative care (Rosa et al. 2022). In LMIC this is still far from being achieved, little progress has been made and somehow suffered setback with the onset of the recent COVID-19 pandemic. Over 60 million people in LMIC have very limited or co access to palliative care (Marie Knaul et al. 2018; Rao et al. 2022). Huge gaps in interventions have been observed in Asia, these include gaps in access to essential medicine, patient management, infrastructure, service planning and accountability. (Palagyi et al. 2019). Improvement of quality of life and wellbeing is still farfetched most LMIC as populations are so diverse socially and economically with the ever-increasing gap between the rich and the poor.

Healthcare access in LMIC has been skewed with poor and vulnerable populations having little and, in some cases, no access even to basic healthcare services. Healthcare access has also been divided between the urban and rural populations, with urban populations having more access to a wide range of healthcare services compared to the rural populations (Hall et al. 2019; Javadi et al. 2020; Tran et al. 2020). Reducing these ever-widening gap and focusing on health needs for the various populations at national, regional and global level and ensuring that the health needs of each population are met could be a step forward towards attaining UHC (Denburg et al. 2019; Hasan et al. 2021).

Achieving UHC and sustaining it requires strong political will and commitment from governments and better health decision making for the benefit of their populations (Hasan et al. 2021). Thus stability in governance in LMIC is required if UHC is to be achieved. In order to sustain good health services there is need for strong financing. This can only be achieved if governments are willing to invest in sustainable healthcare systems with UHC as well as having financial stability to sustain the services (Denburg et al. 2019). Thus if populations within LMIC are able to sustain livelihoods, this could translate to sustainable income, if these income sources are well maintained and managed, these could be invested in creating sustainable health services for the community, population and nation (Domapielle et al. 2022). While progress has been observed in healthcare, setbacks from extreme natural disasters and or sudden disease outbreaks have contributed to challenges in LMIC to attain UHC and maintain sustainable health systems.

3.4.2 Sustained Response to Priority Diseases and COVID-19 Pandemic

Achieving SDG3.3 in light of priority diseases relies on various factors involving animal, human as well as environmental health, social, and economic factors globally. Thus achieving SDG3.3 requires that the rest of the SDG are achieved and sustained. For example as long as there is poverty, lack of access, incquality and climate change, SDG3.3 remains a far-fetched dream. This just shows how complex achieving SDG is and the reason why multisectoral approaches are essential. The COVID-19 pandemic was a great eye opener and was able to reveal not only existing gaps but also policies and strategies that are more nationalistic with minimal global interests (Table 3.6) (Chatterjee et al. 2021). On the other hand, the pandemic also showed the potential of thinking outside the box in terms of new technologies that allowed for faster identification of threats and development of interventions at faster rates than never before. Open access to databases allowed for free flow of information that allowed for identification of new compounds and modelling to identify threats and develop fitting interventions.

Strengthening of global and regional collaborations and encouraging transparency not only in sharing data but sharing information in a timely manner could assist in effective implementation of interventions. Considerations should be taken that there are unknown threats, reservoirs, pathogens, and vectors. All these unknowns create blind spots in attaining SDG3.3; potential crossover and spillover events from these could result in global zoonotic disease threats (Rathish and Vaishnani 2019). Outbreaks from existing pathogens such as fungi and other opportunistic pathogens are worth investigating and prepare for (Brown et al. 2012; Chatterjee et al. 2021; Chowdhary et al. 2017; Hoenigl et al. 2021; Messenger et al. 2014). Thus actively determining the occurrence and diversity of reservoirs and pathogens in various geographical regions is essential although it remains a challenge in LMIC. A study by Breed et al. demonstrates this aspect, occurrence of henipaviruses in fruit bats species (Breed et al. 2013). Creating ways of preventing exposure to reservoirs as well as multi sectoral health approach have been reported as interventions in tackling viral diseases (Rathish and Vaishnani 2019).

Table 3.6 Sustainable health response; successes and threats

Success to Sustainable Response	Threats to Sustainable Response
Ability to devise technical solutions as observed during the COVID-19 pandemic e.g. vaccine and drug development	Threats to global solidarity through actions including vaccine nationalism observed during the pandemic hinder sustainable response in LMIC and globally
Access to advanced technologies such as genome sequencing that allows for prompt and highly specific identification of pathogens	Novel spillover events from threats yet unknown
Access to information including publications and databases	Need for strengthening of regional, global and cross-border collaboration
Access to information from databases and computational approaches which enable modelling and identification of compounds of interest	Limited transparency which cultivate lack of trust hindering interventional programmes
Innovative approaches to risk mitigation	Climate change could be a catalyst of emergence of pathogens in new geographical areas
	Unknown reservoirs and vectors as possible sources of new zoonotic diseases
	New mixing patterns due to globalization and human behavior may result crossovers and potential spillover
	Outbreaks from existing but unsuspected pathogens such as fungi

3.4.2.1 Spillover

For a pathogen to move or spillover from the environment to the human and or animal host and achieve human-to-human transmission is a gradual process. The process involves transitioning of the pathogen from one source to another and adaptation in the host and or environment. To demonstrate this process a pathogen pyramid was used (Wolfe et al. 2007;

Woolhouse and Gowtage-Sequeria 2005). The pathogen pyramid is divided into 4 levels (Fig. 3.3). Level 1 represents exposure, where individuals are exposed to the disease-causing pathogen as they interact with their surroundings. At this point the pathogen could be a known and or completely unknown pathogen. The exposed individual could display traces of immune response from the exposure. Level 2 represents the pathogens' ability to invade human cells. Level 3 is when the pathogen is established in the human body, is able to cause disease and can be transmitted from one person to another, this human-to-human transmission could result in outbreaks. Some examples of level 3 diseases include Ebola virus, Marburg virus disease as well as Nipah virus disease. Level 4 is when the pathogen is established and able to maintain human-to-human transmission through mutations and development of new variants creating a complex cycle of disease. At level 4 the pathogen is capable of causing epidemics and or a pandemic, examples include diseases such as COVID-19, HIV/AIDS with some of the priority diseases listed having greater chances of getting to that level (Chatterjee et al. 2021).

A spillover percolation model reported by Plowright et al., outlined different barriers to zoonotic spillover and opportunities for intervention (Plowright et al. 2017). In CCHFV for example, the determinants of spillover range from the reservoir/vector (the tick), the animal host, human behavior, as well as biological factors and epidemiology in the human host. Ecological dynamics of infection in the tick and animal host influence the carriage and spread of the virus. While Human behavior and epidemiology as well as intrinsic host factors affect exposure and disease outcomes in humans (Plowright et al. 2017). For a spillover to occur, the pathogen has to overcome several barriers with varying levels of constraints. For diseases such as rabies and Ebola virus disease, spillover effects are extremely dangerous and highly fatal due to host response to the disease. In this case if a disease spills over to humans, the host response may not be strong enough to fight the disease. Conversely, in a pathogen such as Toxoplasma gondii, has high exposure but less likely to cause fatal disease due to constraints from intrinsic (within) host barriers. In this case the human host has developed mechanisms that reduce the likelihood of disease spread with fatal outcomes (Plowright et al. 2017).

The likelihood of human infection is thus determined by intrinsic host factors, even though pathogen dose, route of transmission, and human behavior still play a role during human-to-human transmission after spillover.

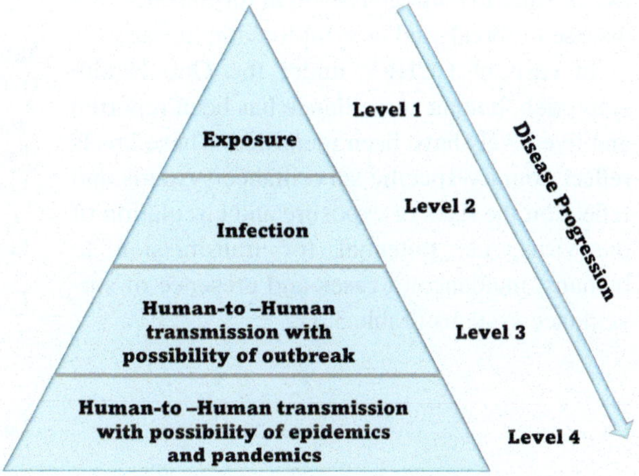

Fig. 3.3 Pyramid showing zoonotic disease progression, from exposure to epidemic and pandemic prone disease

3.4.3 Identifying Gaps in Prevention and Control of Priority Diseases

To prevent incidences of spillover as well as reducing the risks of exposure and spread of zoonotic as well as priority diseases, several strategies have been implemented at national and global level. These include surveillance, risk assessment and risk reduction.

3.4.3.1 Surveillance

Surveillance can be conducted unilaterally within a country and collaboratively across different sectors in the same country and or cross border between neighbouring countries.

For example for diseases that have vectors that span different geographic regions, coordinated surveillance across sectors is a requirement. For zoonotic diseases collaborations between animal and human health sectors is crucial. For a disease such as CCHFV the vector, Hyalomma ticks are found in more than 30 countries in different regions thus surveillance needs to be coordinated between the various countries and regions as well as multiple sectors involved in health (Morens et al. 2004; Taylor et al. 2001).

Depending on the pathogen, different surveillance methods can be used.

Serological and genomic surveillance: This is important in assessing the circulation of the pathogen within the vector, host animals and or humans. This type of surveillance, together with genomic surveillance reveals circulation as well as exposure to pathogens even in the absence of a disease outbreak and or symptomatic disease.

In case of CCHFV under the One Health Approach, human surveillance has been reported and five levels have been identified. These levels reflect country-specific surveillance systems and reflect on the risks of exposure and circulation of the virus, i.e. potential for transmission to humans, incidence of cases and presence of surveillance systems (Table 3.7).

3.4.3.2 Risk Assessment Example of CCHFV

It is crucial that the determinants leading to spillover should be studied not individually but in combination to understand and identify areas that could lead to spillover events (Amman et al. 2017; Goldspink et al. 2015; Hayman et al. 2013; Sorvillo et al. 2020). The interconnection between transmission and vector behavior is of great importance, thus targeting all aspects of transmission is essential to prevent potential spillover (Plowright et al. 2015). It is crucial to understand the interconnectedness of determinants that could lead to spillover events. While these determinants may be dealt with individually, interventions focussed on interconnected parameters could prove more successful.

3.4.3.3 Risk Reduction: Human-Targeted Approaches

Risk reduction is an important aspect of infectious disease prevention. Risks come in many ways including availability of vector, exposure to vector and or disease, human behavior that could perpetuate spread of disease, and the dose and ability of disease pathogen to be passed on from vector to human as well as from human-to-human (Sas et al. 2017; Sorvillo et al. 2020). In human targeted approaches to priority and or zoonotic diseases should also consider occupations that could expose people to the pathogen as well as the vector or any other potentially infectious material such as body fluids and fomites (Surtees

Table 3.7 Surveillance levels for CCHFV (Plowright et al. 2017)

Surveillance Level	Potential Outcomes
Level 1	Countries where CCHFV is endemic, and cases are reported annually
Level 2	Countries with sporadic autochthonous human cases
Level 3 and 4	Countries with no documented human cases but ecologic data, including the presence of Hyalomma ticks, where cases may occur without detection
Level 5	Countries where no information is available

et al. 2016). Management of these physical and chemical barriers have to be carefully mitigated.

Physical and Chemical Measures

For diseases such as CCHF, high risk occupations include veterinarians, abattoir workers, and farmers. Implementing standard disease control practices when handling potentially infectious blood or ticks aids in risk reduction in these work places could be beneficial in reducing exposure (Ergönül 2006).

In Turkey it was observed that approximately 90% of CCHF cases were farmers clearly pointing to associated occupation as one factor that could have contributed to this outcome (Ergönül 2006). Environmental exposure to vector and disease-causing pathogens may not always result in symptomatic disease, but immunological evidence of exposure can be detected in animal and human host (Mediannikov et al. n.d.). The environmental plays a major role in vectorborne diseases, and exposure to vectors such as ticks, mosquitoes, flies, fleas, are linked to setting and geographical location. Thus methods of reducing bites through focus on reducing environmental exposure are crucial in reducing risk of disease transmission and exposure. Physical and chemical methods of reducing exposure such as wearing long sleeves and other appropriate clothing as well as use of chemical repellents to ward off vectors have contributed to reducing transmission (Hoogstraal 1979; Mediannikov et al. n.d.; Sedaghat et al. 2017).

Behavior

Human behavior is unique, in the same country or region, different geographical settings and underlying cultures determine behavior. The interactions between humans and animals are also very different in different cultures and these interactions could increase the risk of exposure in some cultures (Poletti et al. n.d.). In CCHFV practices that allow movement of infected animals have been have been linked to outbreak (Hussain Mallhi et al. 2016; Leblebicioglu et al. 2015).

3.4.3.4 Risk Reduction: Animal-Targeted Approaches

Integrated Control Measures

Zoonotic diseases are an important aspect of public health and human infectious diseases. With over 70% of the emerging diseases in the past decade being of zoonotic origin, animal targeted approached to combat these diseases are crucial (Merianos 2007).

Since zoonotic diseases involve a complex cycle of reservoir host (animal)—vector (arthropod)—host (human) to perpetuate disease, preventative actions have to focus on these aspects and associated determinants. Integrated control strategies are crucial in disease control in wildlife. As such combining knowledge and tools from fields including disease ecology, natural history, pathogen characteristics, domestic and wild animal host characteristics, geographical distribution and human host population, could ensure success (Gortazar et al. 2015).

Some of the measures that have been implemented include, vector control, measures that prevent and or minimize interaction between host and vector, selective culling, management of animal habitats, reproductive control, and vaccinations.

In some zoonotic diseases, the diseases can be transmitted from the wild animal reservoir host to a domesticated animal host which in turn spreads the disease to the human host. Controlling the interface between wild animals and domesticated animals in this case is essential to reduce disease transmission (Gortazar et al. 2015).

One of the control measures that has been used in some instances is vaccination of wild animals, while it is easier to vaccinate domesticated animals, this exercise could prove challenging in wild animals. Some of the barriers that have been reported on wild animal vaccination include involvement of more than one animal species in a disease transmission cycle which affects the decision as to which animal and or animals the vaccine should be targeted to. High turnover of animals due to increased reproductive rate different. Mode of delivery for the vaccine, prevailing environmental conditions and the cost of vaccine production, acquisition and delivery is also another challenge (Monath 2013). Despite these challenges, successful vaccines have been developed for wildlife animals to prevent diseases in humans.

Reservoir-targeted vaccine (RTV) targeting specific animal reservoirs have been suggested as an approach that could be effective in controlling the transmission of some disease pathogens. It has been reported on the potential of RTV to be effective in breaking the transmission cycle of tick-borne pathogens such as, Borrelia burgdorferi, Borrelia miyamotoi, Borrelia mayonii, Babesia microti, and Anaplasma phagocytophilum in mice (Williams et al. 2020).

Transmissible vaccines are those that have the ability to spread from one individual to another. The broad coverage of transmissible vaccines is one essential advantage and equivalent dosing. In order for a vaccine to be categorized as a transmissible vaccine, it must have two essential properties. The vaccine must have a live vector and should be able to spread within the population. Secondly, the vector must remain attenuated during the spread from one vertebrate to the other. Transmission vaccines are thought to be more efficient in increasing herd immunity within a population compared to non-transmissible vaccines (Bull et al. 2018). Despite these advantages consideration should be taken when working with transmissible vaccines. There are chances of reversion which could lead to induction of clinical signs and there are also possibilities of short-lived immunity for certain pathogens.

Limitations of Animal Vaccines

Some of the limitations that have been identified for animal vaccines is that large doses of vaccines may be required in livestock compared to humans. Due to these high dose requirements, production of these vaccines could be costly thereby affecting the production and availability (e.g. DNA, RNA, or virus-like particles) compared to subunit vaccines.

Other diseases that could potentially cause public health emergency include Arenaviral hemorrhagic fevers other than Lassa Fever, Chikungunya, highly pathogenic coronaviral diseases other than MERS and SARS, emergent non-polio enteroviruses (including EV71, D68), Severe Fever with Thrombocytopenia Syndrome (SFTS), Monkeypox and leptospirosis (though not listed as a priority pathogen) have great potential of causing outbreaks.

One silent group of pathogens to look out for are the fungi. Although no major disease outbreak has been reported, increasing drug resistance to existing anti-fungal drugs could result into a super bug capable of causing an epidemic.

All the disease listed above are caused by different pathogens with varying modes of transmission, incubation times, virulence and mortality rates. The question remains can sustained response be achieved to these diseases given the different settings, financial capabilities, and health systems?

3.4.3.5 SDG3d and Emergency Preparedness; Urgent Need for Research and Development on Priority Diseases

It has been observed that diseases affecting low-income populations go unnoticed until the disease becomes a threat to populations in high income countries. One example is mpox disease, previously known as monkeypox diseases. The disease has been known close to 5 decades in endemic countries in Africa but not much attention was paid to it until outbreaks in high income countries occurred. The response to treat and manage the disease in high income countries has been swift with various therapeutics and vaccines being tested to treat the disease. On the other hand, LMIC that have struggled with the disease remain struggling with no access to the vaccines and therapies being used in high income countries.

While a great foundation has been laid for malaria, HIV, and TB prevention and control as well as risk management, preparedness and response globally; such systems are not as developed for most priority diseases. Most of the priority diseases have limited therapeutics while some remain without vaccines or available treatment. As such reducing the risk of acquiring and spread of the disease is of utmost importance. Populations in LMIC are at high risk of emerging and reemerging diseases most of which are capable of causing outbreaks (Akkina et al. 2019; Fischer and Staples 2014; Plowright et al. 2015). Health emergencies and natural disasters have contributed to public health challenges in LMIC, novel pathogens, disease severity and mutations have contributed to complexities in treatment and management of diseases. With limited resources and often minimal and or complete lack of preparedness and risk assessment, most countries in LMIC are struggling with SDG3d. Preparedness is possible if one has ample knowledge of what they are preparing for. Local preparedness is one of the crucial components in sustained response. This includes developing as well as strengthening existing systems so that they can be used in various natural disasters and health emergencies. By creating and maintaining systems at community, provincial and national level and training various stakeholders in management and sustaining the system, using locally available tools could assist in building resilient communities. Limited surveillance and surveillance data in LMIC puts these countries at risk of priority diseases and other health emergencies. Preparedness includes putting together frameworks to manage various types of health emergencies, lack of such puts countries at vulnerable positions as they have no idea where to start from when it comes to responding to these emergencies. Knowledge of what could be required in case of an emergency is available but application and availability of resources to implement the ideas and or framework may vary according to setting. It thus crucial for LMIC to use the available data, knowledge and frameworks to develop strategies that can be implemented within their setting and with the resources available in that setting. The COVID-19 pandemic has revealed gaps in health systems not only globally but also at national level. LMIC should thus use these lessons to plan for better preparedness and response to health emergencies (https://www.who.int/westernpacific/newsroom/feature-stories/item/in-cambodia%2D%2Dwho-s-preparedness-team-recognised-for-helping-build-a-stronger-local-response). These include training personnel in various aspects of managing disease emergencies, acquiring the necessary resources required to respond to an emergency, setting up infrastructure as well as early warning systems to handle emergencies (Herron et al. 2022). Countries should be able to set up priorities and take stock of their capabilities in managing the essential health services while dealing with and after emergencies. Involvement of all stakeholders is crucial in setting up strategies for early warning systems, risk management, preparedness and response to emergencies. While governments have to manage the emergency response, the communities are at the centre of the response, hence dialogue with these communities is crucial in setting up better early warning, preparedness and response strategies that could be easily adopted by the communities to combat global health risks.

3.5 Conclusion

Combating priority diseases and having a sustainable prevention and control strategy is a key factor to achieve the SDG by the year 2030. Since the year 2020, a pandemic of SARS—COVID-19 occurred and its effects are still visible globally. Apart from COVID-19, disease outbreaks from other priority diseases including, Crimean Congo Haemorrhagic fever (CCHF), Ebola virus disease (EBV), Marburg virus disease (MBV), Nipah virus disease (NVD), Rift valley fever (RVF), are still occurring in endemic areas.

There is need to prepare for emerging pathogens with varying characteristics and disease patterns. The preparation should not only look at the health factors but also the social and economic impacts that come with the health interventions aimed at combatting the disease in question. One crucial aspect to ensure that infectious diseases are contained and limit the spread of the disease is contact tracing which is still a challenge both in HIC and LMIC. Several factors have been attributed to this outcome, these include stigma, anonymity (where individuals were unable to identify some of their contacts e.g. monkeypox cases in MSM), as well as limited infrastructure and costs. Improving surveillance and health security in endemic countries as well as LMIC could assist in better management of health emergencies and endemic disease outbreaks. Limited and poor infrastructure was another challenge associated with management of priority diseases in LMIC. Upgrade of infrastructure including building required infrastructure in LMIC and endemic regions is essential to achieve sustainable response. While there is need to improve health security on the African continent, Nigeria's management of the monkeypox outbreak shows systematic response and what can be achieved when there is systematic response. In Bhutan lack of experience and knowledge of Monkeypox was a challenge but the country was able to upgrade infrastructure and create clinical definitions that assisted in improving response. In Latin America, there is a need on working on enhance capacity and creating networks that could assist in data sharing through multisectoral approaches. This could assist in data integration from different sectors and improve management of outbreaks and pandemics. Failures in risk communication, syndemics and poverty were highlighted as some of the challenges that affected preparedness and response in some of the countries in the Latin America region. Strengthening surveillance and provision of PrEP could assist in response to monkeypox. Mathematical modelling is suggested as a way forward to establish efficient systems and expand laboratory capacity including identifying sustainable infection technologies. Improved manufacturing and effective communication are some of the methods that could assist in attaining sustainable preparedness and response in LMIC. Lack of data and poor governance as well as inadequate supply of drugs, vaccines and therapies including registration of these products in LMIC and endemic countries remains a challenge.

References

(CDC), C. for D. C. U. S. of A., Organization, & Health, (WHO) World. (1998) Infection control for viral haemorrhagic fevers in the African health care setting. Centers for disease control and crevention: national center for infectious diseases, division of viral and rickettsial diseases, special pathogens branch, WHO collaborating centre for viral haemorrhagic fevers & world health organization: division of Emer, 198.. http://www.cdc.gov/vhf/abroad/pdf/african-healthcare-setting-vhf.pdf

Aberese-Ako Id M, Dalaba MA, Anumu FEY, Ofosu A, Gyapong M (2022) The socio-economic and health effects of COVID-19 among rural and urban-slum dwellers in Ghana: A mixed methods approach. https://doi.org/10.1371/journal.pone.0271551

Abrahams-Gessel S, Denman CA, Montano CM, Gaziano TA, Levitt N, Rivera-Andrade A et al (2015) Lessons from training and supervision of community health workers conducting non-invasive, population-based screening for cardiovascular disease in four communities in low and middle-income settings: implications for scaling up. Glob Heart 10(1):39. https://doi.org/10.1016/J.GHEART.2014.12.009

Abubakar I, Gram L, Lasoye S, Achiume T, Becares L, Kaur Bola G et al (2022) Confronting the consequences of racism, xenophobia, and discrimination on health and health-care systems. Lancet 400:2137–2146. https://doi.org/10.1016/S0140-6736(22)01989-4

Accorsi S, Fabiani M, Nattabi B, Corrado B, Iriso R, Ayella EO et al (2005) The disease profile of poverty: morbidity and mortality in northern Uganda in the context of war, population displacement and HIV/AIDS. Trans R Soc Trop Med Hyg 99(3):226–233. https://doi.org/10.1016/j.trstmh.2004.09.008

Aceng JR, Ario AR, Muruta AN, Makumbi I, Nanyunja M, Komakech I et al (2020) Uganda's experience in Ebola virus disease outbreak preparedness, 2018-2019. Glob Health 16(1):24. https://doi.org/10.1186/S12992-020-00548-5

Adam EA, Collier SA, Fullerton KE, Gargano JW, Beach MJ (2017) Prevalence and direct costs of emergency department visits and hospitalizations for selected diseases that can be transmitted by water, United States. J Water Health 15(5):673–683. https://doi.org/10.2166/wh.2017.083

Akkina R, Garry R, Bréchot C, Ellerbrok H, Hasegawa H, Menéndez-Arias L et al (2019) 2019 meeting of the global virus network. Antivir Res 172:104645. https://doi.org/10.1016/j.antiviral.2019.104645

Akseer N, Kandru G, Keats EC, Bhutta ZA (2020) COVID-19 pandemic and mitigation strategies: implications for maternal and child health and nutrition. Am J Clin Nutr 112(2):251–256. https://doi.org/10.1093/ajcn/nqaa171

Al Awaidy S, Al Hashami H (2020) Zoonotic diseases in Oman: successes, challenges, and future directions.

Vector Borne Zoonotic Dis 20(1):1–9. https://doi.org/10.1089/vbz.2019.2458

Alhaji NB, Babalobi OO, Wungak Y, Ularamu HG (2018) Participatory survey of Rift Valley fever in nomadic pastoral communities of North-Central Nigeria: the associated risk pathways and factors. PLoS Negl Trop Dis 12(10):e0006858. https://doi.org/10.1371/journal.pntd.0006858

Amman BR, Swanepoel R, Nichol ST, Towner JS (2017) Ecology of filoviruses. Curr Top Microbiol Immunol 411:23–61. https://doi.org/10.1007/82_2017_10

Aro T, Kantele A (2018) Among asylum seekers and refugees admitted to Helsinki university hospital, 2010 to 2017. Euro Surveill 23(45):1700797

Assefa Y, Gilks CF (2020) Ending the epidemic of HIV/AIDS by 2030: will there be an endgame to HIV, or an endemic HIV requiring an integrated health systems response in many countries? Int J Infect Dis 100:273–277. https://doi.org/10.1016/J.IJID.2020.09.011

Awah PK, Boock AU, Mou F, Koin JT, Anye EM, Noumen D, Nichter M (2018) Developing a Buruli ulcer community of practice in Bankim, Cameroon: a model for Buruli ulcer outreach in Africa. PLoS Negl Trop Dis 12(3):e0006238. https://doi.org/10.1371/journal.pntd.0006238

Beattie TS, Bhattacharjee P, Isac S, Davey C, Javalkar P, Nair S et al (2015) Supporting adolescent girls to stay in school, reduce child marriage and reduce entry into sex work as HIV risk prevention in North Karnataka, India: protocol for a cluster randomised controlled trial. BMC Public Health 15:292. https://doi.org/10.1186/s12889-015-1623-7

Belkina TV, Khojiev DS, Tillyashaykhov MN, Tigay ZN, Kudenov MU, Tebbens JD, Vlcek J (2014) Delay in the diagnosis and treatment of pulmonary tuberculosis in Uzbekistan: a cross-sectional study. BMC Infect Dis 14:624. https://doi.org/10.1186/s12879-014-0624-y

Bell BP, Boyle CA, Petersen LR (2016, April) Preventing zika virus infections in pregnant women: an urgent public health priority. Am J Public Health 106:589–590. https://doi.org/10.2105/AJPH.2016.303124

Benedict K, Jackson BR, Chiller T, Beer KD (2019) Estimation of direct healthcare costs of fungal diseases in the United States. Clin Infect Dis 68(11):1791–1797. https://doi.org/10.1093/cid/ciy776

Bisika T (2010) Health systems strengthening in conflict situations. East Afr J Public Health 7(3):277–281. http://www.ncbi.nlm.nih.gov/pubmed/21651793

Bonwitt J, Dawson M, Kandeh M, Ansumana R, Sahr F, Brown H, Kelly AH (2018) Unintended consequences of the "bushmeat ban" in West Africa during the 2013-2016 Ebola virus disease epidemic. Soc Sci Med 200:166–173. https://doi.org/10.1016/j.socscimed.2017.12.028

Borremans B, Faust C, Manlove KR, Sokolow SH, Lloyd-Smith JO (2019) Cross-species pathogen spillover across ecosystem boundaries: mechanisms and theory. Philos Trans R Soc Lond Ser B Biol Sci 374(1782):20180344. https://doi.org/10.1098/rstb.2018.0344

Boulware LE, Cooper LA, Ratner LE, LaVeist TA, Powe NR (2003) Race and trust in the health care system. Public Health Rep 118(4):358. https://doi.org/10.1016/S0033-3549(04)50262-5

Bourgarel M, Wauquier N, Gonzalez J-P (2010) Emerging viral threats in Gabon: health capacities and response to the risk of emerging zoonotic diseases in Central Africa. Emerg Health Threats J 3:e7. https://doi.org/10.3134/ehtj.10.163

Bower H, El Karsany M, Alzain M, Gannon B, Mohamed R, Mahmoud I et al (2019) Detection of Crimean-Congo haemorrhagic fever cases in a severe undifferentiated febrile illness outbreak in the federal republic of Sudan: a retrospective epidemiological and diagnostic cohort study. PLoS Negl Trop Dis 13(7):e0007571. https://doi.org/10.1371/journal.pntd.0007571

Breed AC, Meers J, Sendow I, Bossart KN, Barr JA, Smith I et al (2013) The distribution of henipaviruses in Southeast Asia and Australasia: is Wallace's line a barrier to Nipah virus? PLoS One 8(4):61316. https://doi.org/10.1371/journal.pone.0061316

Brotherhood L, Cavalcanti T, Da Mata D, Santos C (2022) Slums and pandemics. J Dev Econ 157:102882. https://doi.org/10.1016/j.jdeveco.2022.102882

Brown GD, Denning DW, Gow NAR, Levitz SM, Netea MG, White TC (2012) Hidden killers: human fungal infections. Sci Transl Med 4:165. https://doi.org/10.1126/scitranslmed.3004404

Buheji M, da Costa Cunha K, Beka G, Mavrić B, Leandro do Carmo de Souza Y, Souza da Costa Silva S et al (2020) The extent of COVID-19 pandemic socio-economic impact on global poverty. A global integrative multidisciplinary review. Am J Econ 10(4):213–224. https://doi.org/10.5923/j.economics.20201004.02

Bull JJ, Smithson MW, Nuismer SL (2018) Transmissible viral vaccines. Trends Microbiol 26(1):6–15. https://doi.org/10.1016/J.TIM.2017.09.007

Bundervoet T, Dávalos ME, Garcia N (2021) The short-term impacts of COVID-19 on households in developing countries. World bank policy research working paper 9582, (March)

Burkle FM (2020) Political intrusions into the international health regulations treaty and its impact on management of rapidly emerging zoonotic pandemics: what history tells us. Prehosp Disaster Med 35(4):426–430. https://doi.org/10.1017/S1049023X20000515

Burkle FM Jr (2017) The politics of global public health in fragile states and ungoverned territories. PLoS Curr 9:ecurrents.dis.ba3beede71ca0746a0972aa3837ed618. https://doi.org/10.1371/CURRENTS.DIS.BA3BEEDE71CA0746A0972AA3837ED618

Caballero-Anthony M (2021) COVID-19 in Southeast Asia: regional pandemic preparedness matters. Order from Chaos [BLOG]. http://libaccess.mcmaster.ca/login?url=https://www.proquest.com/blogs-podcasts-websites/covid-19-southeast-asia-regional-pandemic/docview/2477799720/se-2?accountid=12347%0Ahttp://sfx.

scholarsportal.info/mcmaster?url_ver=Z39.88-2004&rft_val_fmt=info:ofi/fm

Cauchemez S, Epperson S, Biggerstaff M, Swerdlow D, Finelli L, Ferguson NM (2013) Using routine surveillance data to estimate the epidemic potential of emerging zoonoses: application to the emergence of US swine origin influenza a H3N2v virus. PLoS Med 10(3):e1001399. https://doi.org/10.1371/journal.pmed.1001399

Chalem E, Mitsuhiro SS, Ferri CP, Barros MCM, Guinsburg R, Laranjeira R (2007) Teenage pregnancy: behavioral and socio-demographic profile of an urban Brazilian population. Cad Saude Publica 23(1):177–186. https://doi.org/10.1590/s0102-311x2007000100019

Chatterjee P, Nair P, Chersich M, Terefe Y, Chauhan AS, Quesada F, Simpson G (2021) One health, "disease X" & the challenge of "unknown" unknowns. Indian J Med Res 153:264–271. https://doi.org/10.4103/ijmr.IJMR_601_21

Chauhan RP, Dessie ZG, Noreddin A, El Zowalaty ME (2020) Systematic review of important viral diseases in Africa in light of the "one health" concept. Pathogens 9(4):301. https://doi.org/10.3390/pathogens9040301

Chowdhary A, Sharma C, Meis JF (2017) Candida auris: a rapidly emerging cause of hospital-acquired multidrug-resistant fungal infections globally. https://doi.org/10.1371/journal.ppat.1006290

Chua AQ, Tan MMJ, Verma M, Han EKL, Hsu LY, Cook AR et al (2020) Health system resilience in managing the COVID-19 pandemic: lessons from Singapore. BMJ Glob Health 5(9):e003317. https://doi.org/10.1136/bmjgh-2020-003317

Clark J, Stolk WA, Basáñez M-G, Coffeng LE, Cucunubá ZM, Dixon MA et al (2021) How modelling can help steer the course set by the world health organization 2021-2030 roadmap on neglected tropical diseases. Gates Open Res 5:112. https://doi.org/10.12688/gatesopenres.13327.1

Contini C, Caselli E, Martini F, Maritati M, Torreggiani E, Seraceni S et al (2020) COVID-19 is a multifaceted challenging pandemic which needs urgent public health interventions. Microorganisms 8(8):1228. https://doi.org/10.3390/microorganisms8081228

Cowling BJ, Jin L, Lau EHY, Liao Q, Wu P, Jiang H et al (2013) Comparative epidemiology of human infections with avian influenza A H7N9 and H5N1 viruses in China: a population-based study of laboratory-confirmed cases. Lancet 382(9887):129–137. https://doi.org/10.1016/S0140-6736(13)61171-X

Das D, Baker M, Venugopal K, McAllister S (2006) Why the tuberculosis incidence rate is not falling in New Zealand. N Z Med J 119(1243):U2248

Das M, Angeli F, Krumeich AJSM, Van Schayck OCP (2018) Patterns of illness disclosure among Indian slum dwellers: A qualitative study. BMC Int Health Hum Rights 18:3. https://doi.org/10.1186/s12914-018-0142-x

Davey G, Burridge E (2009) Community-based control of a neglected tropical disease: the mossy foot treatment and prevention association. PLoS Negl Trop Dis 3(5):e424. https://doi.org/10.1371/journal.pntd.0000424

De Pauw B, Walsh TJ, Donnelly JP, Stevens DA, Edwards JE, Calandra T et al (2008) Revised definitions of invasive fungal disease from the European organization for research and treatment of cancer/invasive fungal infections cooperative group and the national institute of allergy and infectious diseases mycoses study group (EORTC/MSG) C. Clin Infect Dis 46(12):1813–1821. https://doi.org/10.1086/588660

Deml MJ, Githaiga JN (2022) Original research: determinants of COVID-19 vaccine hesitancy and uptake in sub-Saharan Africa: a scoping review. BMJ Open 12(11):e066615. https://doi.org/10.1136/BMJOPEN-2022-066615

Denburg AE, Ramirez A, Pavuluri S, McCann E, Shah S, Alcasabas T et al (2019) Political priority and pathways to scale-up of childhood cancer care in five nations. PLoS One 14(8):e0221292. https://doi.org/10.1371/JOURNAL.PONE.0221292

Devakumar D, Selvarajah S, Abubakar I, Kim S-S, McKee M, Sabharwal NS et al (2022a) Racism, xenophobia, discrimination, and the determination of health. Lancet 400(10368):2097–2108. https://doi.org/10.1016/S0140-6736(22)01972-9

Devakumar D, Selvarajah S, Abubakar I, Kim S-S, Mckee M, Sabharwal NS et al (2022b) Racism, xenophobia, discrimination, and health 1 racism, xenophobia, discrimination, and the determination of health. Lancet 400:2097–2108. https://doi.org/10.1016/S0140-6736(22)01972-9

Dionne KY, Turkmen FF (n.d.) The politics of pandemic othering: putting COVID-19 in global and historical context, vol 74. International Organization, pp E213–E230. https://doi.org/10.1017/S0020818320000405

Dixon MG, Schafer IJ (2014) Ebola viral disease outbreak--West Africa, 2014. MMWR Morb Mortal Wkly Rep 63(25):548–551. https://doi.org/10.1016/j.annemergmed.2014.10.010

Domapielle MK, Sumankuuro J, Der Bebelleh F (2022) Revisiting the debate on health financing in low and middle-income countries: an integrative review of selected models. Int J Health Plann Manag 37(6):3061–3074. https://doi.org/10.1002/HPM.3566

Dungu B, Lubisi BA, Ikegami T (2018) Rift Valley fever vaccines: current and future needs. Curr Opin Virol 29:8–15. https://doi.org/10.1016/j.coviro.2018.02.001

Ellington S, Perez M, Morof D, Zotti ME, Callaghan W, Meaney-Delman D et al (2017) Addressing maternal health during CDC's Ebola response in the United States. J Womens Health (Larchmt) 26(11):1141–1145. https://doi.org/10.1089/jwh.2017.6719

Elphick-Pooley T, Engels D (2020) Improvement cues of lesion absorption using the adjuvant therapy of traditional Chinese medicine Qinbudan tablet for retreatment pulmonary tuberculosis with standard

anti-tuberculosis regimen. https://doi.org/10.1186/s40249-021-00932-2
Engels D, Zhou XN (2020) Neglected tropical diseases: an effective global response to local poverty-related disease priorities. Infect Dis Poverty 9(1):1–9. https://doi.org/10.1186/s40249-020-0630-9
Ergönül Ö (2006) Crimean-Congo haemorrhagic fever. Lancet Infect Dis 6:203–214. https://doi.org/10.1016/S1473-3099(06)70435-2
Farag E, Sikkema R, Vinks T, Islam M, Nour M, Al-Romaihi H et al (2018) Drivers of MERS-CoV emergence in Qatar. Viruses 11(1):22. https://doi.org/10.3390/v11010022
Farmer PE (2013) Chronic infectious disease and the future of health care delivery. N Engl J Med 369(25):2424–2436. https://doi.org/10.1056/NEJMsa1310472
Feldmann H, Feldmann F, Marzi A (2018) Ebola: Lessons on Vaccine Development. Annu Rev Microbiol 72:423–446. https://doi.org/10.1146/annurev-micro-090817-062414
Fillâtre P, Revest M, Tattevin P (2019) Crimean-Congo hemorrhagic fever: an update. Med Mal Infect 49(8):574–585. https://doi.org/10.1016/J.MEDMAL.2019.09.005
Fischer M, Staples JE (2014) Notes from the field: chikungunya virus spreads in the Americas-Caribbean and South America, 2013-2014. MMWR Morb Mortal Wkly Rep 63(22):500–501
Gambhir M, Bozio C, O'Hagan JJ, Uzicanin A, Johnson LE, Biggerstaff M, Swerdlow DL (2015) Infectious disease modeling methods as tools for informing response to novel influenza viruses of unknown pandemic potential. Clin Infect Dis 1(Suppl 1):S11–S19. https://doi.org/10.1093/cid/civ083
Gayer M, Legros D, Formenty P, Connolly MA (2007) Conflict and emerging infectious diseases. Emerg Infect Dis 13:1625–1631. https://doi.org/10.3201/eid1311.061093
Gell JS (2015) Alternative medicine. In: Clinical gynecology, 2nd edn. https://doi.org/10.1017/CBO9781139628938.008
George LS, Rakesh PS, Vijayakumar K, Kunoor A, Kumar A (2020) Social stigma associated with TB and HIV/AIDS among Kudumbashree members: a cross-sectional study. J Family Med Prim Care 9(8):4062. https://doi.org/10.4103/JFMPC.JFMPC_437_20
Gilson L, Lehmann U, Schneider H (2017) Practicing governance towards equity in health systems: LMIC perspectives and experience Lucy Gilson. Int J Equity Health 16(1):1–5. https://doi.org/10.1186/s12939-017-0665-0
Goldspink LK, Edson DW, Vidgen ME, Bingham J, Field HE, Smith CS (2015) Natural hendra virus infection in flying-foxes-tissue tropism and risk factors. PLoS One 10(6):e0128835. https://doi.org/10.1371/journal.pone.0128835
Golechha M (2020) COVID-19, India, lockdown and psychosocial challenges: what next? Int J Soc Psychiatry 66(8):830. https://doi.org/10.1177/0020764020935922
Gómez Román R, Wang L-F, Lee B, Halpin K, de Wit E, Broder CC et al (2020) Nipah@20: lessons learned from another virus with pandemic potential. MSphere 5:4. https://doi.org/10.1128/mSphere.00602-20
Gonzalez JP, Camicas JL, Cornet JP, Faye O, Wilson ML (1992) Sexual and transovarian transmission of Crimean-Congo haemorrhagic fever virus in Hyalomma truncatum ticks. Res Virol 143(1):23–28. https://doi.org/10.1016/S0923-2516(06)80073-7
Gortazar C, Diez-Delgado I, Barasona JA, Vicente J, De J, Fuente L et al (2015) The wild side of disease control at the wildlife-livestock-human interface: a review. Front Vet Sci 1:27. https://doi.org/10.3389/fvets.2014.00027
Gray GC, Kayali G (2009) Facing pandemic influenza threats: the importance of including poultry and swine workers in preparedness plans. Poult Sci 88(4):880–884. https://doi.org/10.3382/ps.2008-00335
Griffin DE (2010) Emergence and re-emergence of viral diseases of the central nervous system. Prog Neurobiol 91(2):95–101. https://doi.org/10.1016/j.pneurobio.2009.12.003
Hall T, Kakuma R, Palmer L, Martins J, Minas H, Kermode M (2019) Are people-centred mental health services acceptable and feasible in timor-leste? A qualitative study. Health Policy Plan 34(Supplement_2):II93–II103. https://doi.org/10.1093/HEAPOL/CZZ108
Hallam HJ, Hallam S, Rodriguez SE, Barrett ADT, Beasley DWC, Chua A et al (2018) Baseline mapping of lassa fever virology, epidemiology and vaccine research and development. NPJ Vaccines 3:11. https://doi.org/10.1038/s41541-018-0049-5
Haq Z, Mahjour J, Khan W (2013) Communicable diseases in the Eastern Mediterranean region: prevention and control 2010–2011. East Mediterr Health J 19(10):888–891
Haroun Y, Sambaiga R, Sarkar N, Kapologwe NA, Kengia J, Liana J et al (2022) A human centred approach to digital technologies in health care delivery among mothers, children and adolescents. BMC Health Serv Res 22(1):1–15. https://doi.org/10.1186/s12913-022-08744-2
Hartley D (2004) Rural health disparities, population health, and rural culture. Am J Public Health 94(10):1675–1678
Hasan BS, Rasheed MA, Wahid A, Kumar RK, Zuhlke L (2021) Generating evidence from contextual clinical research in low- to middle income countries: a roadmap based on theory of change. Front Pediatr 9:764239. https://doi.org/10.3389/FPED.2021.764239
Hayman DTS, Bowen RA, Cryan PM, McCracken GF, O'Shea TJ, Peel AJ et al (2013) Ecology of zoonotic infectious diseases in bats: current knowledge and future directions. Zoonoses Public Health 60(1):2–21. https://doi.org/10.1111/zph.12000
Herron LM, Phillips G, Brolan CE, Mitchell R, O'Reilly G, Sharma D et al (2022) "When all else fails you have to come to the emergency department": overarching lessons about emergency care resilience from frontline clinicians in Pacific Island countries and ter-

ritories during the COVID-19 pandemic. Lancet Reg Health West Pac 25:100519. https://doi.org/10.1016/J.LANWPC.2022.100519

Hoenigl M, Salmanton-García J, Walsh TJ, Nucci M, Neoh CF, Jenks JD et al (2021) Global guideline for the diagnosis and management of rare mould infections: an initiative of the European confederation of medical mycology in cooperation with the international society for human and animal mycology and the American society for microbiology. Lancet Infect Dis 0(0):1–12. https://doi.org/10.1016/S1473-3099(20)30784-2

Hong G, Miller HB, Allgood S, Lee R, Lechtzin N, Zhang SX (2017) Use of selective fungal culture media increases rates of detection of fungi in the respiratory tract of cystic fibrosis patients. J Clin Microbiol 55(4):1122–1130. https://doi.org/10.1128/JCM.02182-16

Hoogstraal H (1979) Review article: the epidemiology of tick-borne Crimean-Congo hemorrhagic fever in Asia, Europe, and Africa. J Med Entomol 15(4):307–417. https://doi.org/10.1093/JMEDENT/15.4.307

Hotez PJ, Savioli L, Fenwick A (2012) Neglected tropical diseases of the Middle East and North Africa: review of their prevalence, distribution, and opportunities for control. PLoS Negl Trop Dis 6(2):e1475. https://doi.org/10.1371/journal.pntd.0001475

Hotez PJ, Bottazzi ME, Strych U, Chang LY, Lim YAL, Goodenow MM, AbuBakar S (2015) Neglected tropical diseases among the Association of Southeast Asian Nations (ASEAN): overview and update. PLoS Negl Trop Dis 9(4):e0003575. https://doi.org/10.1371/journal.pntd.0003575

Hussain Mallhi T, Habib Khan Y, Sarriff A, Hayat Khan A (2016) Crimean-Congo haemorrhagic fever virus and Eid-Ul-Adha festival in Pakistan. https://doi.org/10.1016/S1473-3099(16)30453-4

Hyder S, Sievers BL, Flamand C, TagoPacheco D, Chan M, Claes F, Karlsson EA (2023) Influx of backyard farming with limited biosecurity due to the COVID-19 pandemic carries an increased risk of zoonotic spillover in Cambodia. Microbiol Spectr 11(1):e0420722. https://doi.org/10.1128/spectrum.04207-22

Iglesias Martínez E, Roces García J, Jiménez Arberas E, Llosa JA (2022) Difference between impacts of COVID-19 on women and Men's psychological, social, vulnerable work situations, and economic well-being. Int J Environ Res Public Health 19(14):8849. https://doi.org/10.3390/ijerph19148849

Jajoo M, Manchanda V, Chaurasia S, Jeeva Sankar M, Gautam H, Agarwal R et al (2018) Alarming rates of antimicrobial resistance and fungal sepsis in outborn neonates in North India. https://doi.org/10.1371/journal.pone.0180705

Jamieson DJ, Uyeki TM, Callaghan WM, Meaney-Delman D, Rasmussen SA (2014) What obstetrician-gynecologists should know about Ebola: a perspective from the centers for disease control and prevention. Obstet Gynecol 124(5):1005–1010. https://doi.org/10.1097/AOG.0000000000000533

Javadi D, Ssempebwa J, Isunju JB, Yevoo L, Amu A, Nabiwemba E et al (2020) Implementation research on sustainable electrification of rural primary care facilities in Ghana and Uganda. Health Policy Plan 35(Supplement_2):II124–II136. https://doi.org/10.1093/HEAPOL/CZAA077

Jayatilleke K (2020) Challenges in implementing surveillance tools of high-income countries (HICs) in low middle income countries (LMICs). Curr Treat Options Infect Dis 12(3):191–201. https://doi.org/10.1007/s40506-020-00229-2

Joint United Nations Programme on HIV/AIDS. (2022). UNAIDS global AIDS update 2022. UNAIDS global AIDS update 2022, 2022, 1–22

Jourdain F, Samy AM, Hamidi A, Bouattour A, Alten B, Faraj C et al (2019) Towards harmonisation of entomological surveillance in the Mediterranean area. PLoS Negl Trop Dis 13(6):e0007314. https://doi.org/10.1371/journal.pntd.0007314

Jowell A, Barry M (2020) COVID-19: A matter of planetary, not only national health. Am J Trop Med Hygiene 103(1):31–32. https://doi.org/10.4269/ajtmh.20-0419

Kabuga AI, El Zowalaty ME (2019) A review of the monkeypox virus and a recent outbreak of skin rash disease in Nigeria. J Med Virol 91(4):533–540. https://doi.org/10.1002/jmv.25348

Kamper-Jørgensen Z, Andersen AB, Kok-Jensen A, Kamper-Jørgensen M, Bygbjerg IC, Andersen PH et al (2012) Migrant tuberculosis: the extent of transmission in a low burden country. BMC Infect Dis 12(1):60. https://doi.org/10.1186/1471-2334-12-60

Kasi KK, von Arnim F, Schulz A, Rehman A, Chudhary A, Oneeb M et al (2020) Crimean-Congo haemorrhagic fever virus in ticks collected from livestock in Balochistan, Pakistan. Transbound Emerg Dis 67(4):1543–1552. https://doi.org/10.1111/tbed.13488

Kelly TR, Karesh WB, Johnson CK, Gilardi KVK, Anthony SJ, Goldstein T et al (2017) One health proof of concept: bringing a transdisciplinary approach to surveillance for zoonotic viruses at the human-wild animal interface. Prev Vet Med 137(Pt B):112–118. https://doi.org/10.1016/j.prevetmed.2016.11.023

Kelly-Hope LA, Sanders AM, Harding-Esch E, Willems J, Ahmed F, Vincer F, Hill R (n.d.) Complex emergencies and the control and elimination of neglected tropical diseases in Africa: developing a practical approach for implementing safe and effective mapping and intervention strategies. Conflict and Health. 15:18. https://doi.org/10.1186/s13031-021-00356-7

Khan T, Qazi J (2013) Hurdles to the global antipolio campaign in Pakistan: an outline of the current status and future prospects to achieve a polio free world. J Epidemiol Community Health 67(8):696–702. https://doi.org/10.1136/jech-2012-202162

Khankeh H, Farrokhi M, Roudini J, Pourvakhshoori N, Ahmadi S, Abbasabadi-Arab M et al (2021) Challenges to manage pandemic of coronavirus disease (COVID-19) in Iran with a special situation: a qualitative multi-method study. BMC Public Health 21(1):1–9. https://doi.org/10.1186/s12889-021-11973-5

Knight SR, Ho A, Pius R, Buchan I, Carson G, Drake TM et al (2020) Risk stratification of patients admitted to hospital with covid-19 using the ISARIC WHO clinical characterisation protocol: development and validation of the 4C mortality score. BMJ (Clinical Research Ed) 370:m3339. https://doi.org/10.1136/bmj.m3339

Kollie K, Siakeh A, Zawolo G, Wickenden A, Theobald S, Rogers E et al (2020) Donor reliance and the impact on neglected tropical disease programme delivery: reflections and solutions for change from programme management perspectives. Int Health 0:1–3. https://doi.org/10.1093/inthealth/ihaa090

Kuruvilla S, Hinton R, Boerma T, Bunney R, Casamitjana N, Cortez R et al (2018) Business not as usual: how multisectoral collaboration can promote transformative change for health and sustainable development. BMJ (Online) 363:k4771. https://doi.org/10.1136/bmj.k4771

Leblebicioglu H, Sunbul M, Memish ZA, Al-Tawfiq JA, Bodur H, Ozkul A et al (2015) Consensus report: preventive measures for Crimean-Congo hemorrhagic fever during Eid-al-Adha festival. Int J Infect Dis 38:9–15. https://doi.org/10.1016/j.ijid.2015.06.029

Leblebicioglu H, Ozaras R, Fletcher TE, Beeching NJ (2016) Crimean-Congo haemorrhagic fever in travellers: a systematic review. Travel Med Infect Dis 14(2):73–80. https://doi.org/10.1016/j.tmaid.2016.03.002

Leclerc H, Schwartzbrod L, Dei-Cas E (2002) Microbial agents associated with waterborne diseases. Crit Rev Microbiol 28(4):371–409. https://doi.org/10.1080/1040-840291046768

Loyse A, Burry J, Cohn J, Ford N, Chiller T, Ribeiro I et al (2019) Leave no one behind: response to new evidence and guidelines for the management of cryptococcal meningitis in low-income and middle-income countries. Lancet Infect Dis 19(4):e143–e147. https://doi.org/10.1016/S1473-3099(18)30493-6

Lukoye D, Adatu F, Musisi K, Kasule GW, Were W, Odeke R et al (2013) Anti-tuberculosis drug resistance among new and previously treated sputum smear-positive tuberculosis patients in Uganda: results of the first national survey. PLoS One 8(8):e70763. https://doi.org/10.1371/journal.pone.0070763

MacPherson DW, Gushulak BD, Baine WB, Bala S, Gubbins PO, Holtom P, Segarra-Newnham M (2009) Population mobility, globalization, and antimicrobial drug resistance. Emerg Infect Dis 15(11):1727–1732. https://doi.org/10.3201/eid1511.090419

Maman I, Tchacondo T, Kere AB, Piten E, Beissner M, Kobara Y et al (2018) Risk factors for mycobacterium ulcerans infection (Buruli ulcer) in Togo - a case-control study in Zio and Yoto districts of the maritime region. BMC Infect Dis 18(1):48. https://doi.org/10.1186/s12879-018-2958-3

Marie Knaul F, Farmer PE, Krakauer EL, De Lima L, Bhadelia A, Jiang Kwete X et al (2018) The lancet commissions alleviating the access abyss in palliative care and pain relief-an imperative of universal health coverage: the lancet commission report. Lancet 391:1391–1454. https://doi.org/10.1016/S0140-6736(17)32513-8

Marks M, Mitjà O, Solomon AW, Asiedu KB, Mabey DC (2014) Yaws. Br Med Bull 113:91–100. https://doi.org/10.1093/bmb/ldu037

McClarty L, Lazarus L, Pavlova D, Reza-Paul S, Balakireva O, Kimani J et al (2022) Socioeconomic burdens of the COVID-19 pandemic on LMIC populations with increased HIV vulnerabilities. Curr HIV/AIDS Rep 19:76–85. https://doi.org/10.1007/s11904-021-00591-w

Mead PS, Hills SL, Brooks JT (2018) Zika virus as a sexually transmitted pathogen. Curr Opin Infect Dis 31(1):39–44. https://doi.org/10.1097/QCO.0000000000000414

Mediannikov O, Diatta G, Fenollar F, Sokhna C, Trape J-F, Raoult D (n.d.) Tick-borne rickettsioses, neglected emerging diseases in rural Senegal. PLoS Negl Trop Dis. 4(9):e821. https://doi.org/10.1371/journal.pntd.0000821

Meltzer MI, Santibanez S, Fischer LS, Merlin TL, Adhikari BB, Atkins CY et al (2016) Modeling in real time during the Ebola response. MMWR Suppl 65(3):85–89. https://doi.org/10.15585/mmwr.su6503a12

Merianos A (2007) Surveillance and response to disease emergence. Curr Top Microbiol Immunol 315:477–509. https://doi.org/10.1007/978-3-540-70962-6_19

Mertens M, Schmidt K, Ozkul A, Groschup MH (2013) The impact of Crimean-Congo hemorrhagic fever virus on public health. Antivir Res 98(2):248–260. https://doi.org/10.1016/j.antiviral.2013.02.007

Messenger AM, Barnes AN, Gray GC (2014) Reverse zoonotic disease transmission (zooanthroponosis): a systematic review of seldom-documented human biological threats to animals. PLoS One 9(2):e89055. https://doi.org/10.1371/journal.pone.0089055

Monath TP (2013) Vaccines against diseases transmitted from animals to humans: a one health paradigm. Vaccine 31(46):5321–5338. https://doi.org/10.1016/j.vaccine.2013.09.029

Morens DM, Folkers GK, Fauci AS (2004) The challenge of emerging andre-emerging infectious diseases. Nature 430:242–249. www.nature.com/nature

Mostafavi E, Pourhossein B, Esmaeili S, Bagheri Amiri F, Khakifirouz S, Shah-Hosseini N, Tabatabaei SM (2017) Seroepidemiology and risk factors of Crimean-Congo hemorrhagic fever among butchers and slaughterhouse workers in Southeastern Iran. Int J Infect Dis 64:85–89. https://doi.org/10.1016/j.ijid.2017.09.008

Mphande FA (2016a) Infectious diseases and rural livelihood in developing countries. https://doi.org/10.1007/978-981-10-0428-5

Mphande FA (2016b) Poverty, health and livelihoods. In: Infectious diseases and rural livelihoods in developing countries, pp 1–16. https://doi.org/10.1007/978-981-10-0428-5

Mubarak MY, Wagner AL, Asami M, Carlson BF, Boulton ML (2016) Hygienic practices and diarrheal illness among persons living in at-risk settings in Kabul, Afghanistan: a cross-sectional study.

BMC Infect Dis. 16(1):459. https://doi.org/10.1186/s12879-016-1789-3
Mukherjee S (2020) Disparities, desperation, and divisiveness: coping with COVID-19 in India. Psychol Trauma 12(6):582. https://doi.org/10.1037/TRA0000682
Mutembo S, Mutanga JN, Musokotwane K, Kanene C, Dobbin K, Yao X et al (2019) Urban-rural disparities in treatment outcomes among recurrent TB cases in Southern province, Zambia. BMC Infect Dis 19(1):1087. https://doi.org/10.1186/s12879-019-4709-5
Nguyen MH, Wissel MC, Shields RK, Salomoni MA, Hao B, Press EG et al (2012) Performance of candida real-time polymerase chain reaction, β-D-glucan assay, and blood cultures in the diagnosis of invasive candidiasis. Clin Infect Dis 54(9):1240–1248. https://doi.org/10.1093/cid/cis200
Nuwematsiko R, Nabiryo M, Bomboka JB, Nalinya S, Musoke D, Okello D, Wanyenze RK (2022) Unintended socio-economic and health consequences of COVID-19 among slum dwellers in Kampala, Uganda. BMC Public Health 22(1):1–13. https://doi.org/10.1186/S12889-021-12453-6/FIGURES/5
Oramasionwu CU, Daniels KR, Labreche MJ, Frei CR (2011) The environmental and social influences of HIV/AIDS in sub-Saharan Africa: a focus on rural communities. Int J Environ Res Public Health 8(7):2967–2979. https://doi.org/10.3390/ijerph8072967
Palagyi A, Dodd R, Jan S, Nambiar D, Joshi R, Tian M et al (2019) Organisation of primary health care in the Asia-Pacific region: developing a prioritised research agenda analysis. BMJ Glob Health 4:1467. https://doi.org/10.1136/bmjgh-2019-001467
Papa A, Mirazimi A, Köksal I, Estrada-Pena A, Feldmann H (2015) Recent advances in research on Crimean-Congo hemorrhagic fever. J Clin Virol 64:137–143. https://doi.org/10.1016/j.jcv.2014.08.029
Pascall DJ, Nomikou K, Bréard E, Zientara S, Filipe A d S, Hoffmann B et al (2020) "Frozen evolution" of an RNA virus suggests accidental release as a potential cause of arbovirus re-emergence. PLoS Biol 18(4):e3000673. https://doi.org/10.1371/journal.pbio.3000673
Pereira M, Oliveira AM (n.d.) Commentary poverty and food insecurity may increase as the threat of COVID-19 spreads. Public Health Nutr 23(17):3236–3240. https://doi.org/10.1017/S1368980020003493
Perry HB, Zulliger R, Rogers MM (2014) Community health Workers in low-, middle-, and high-income countries: an overview of their history, recent evolution, and current effectiveness. Annu Rev Public Health 35(1):399–421. https://doi.org/10.1146/annurev-publhealth-032013-182354
Petersen LR, Beard CB, Visser SN (2019) Combatting the increasing threat of vector-borne disease in the united states with a national vector-borne disease prevention and control system. Am J Trop Med Hyg 100(2):242–245. https://doi.org/10.4269/ajtmh.18-0841
Pigott DM, Deshpande A, Letourneau I, Morozoff C, Reiner RCJ, Kraemer MUG et al (2017) Local, national, and regional viral haemorrhagic fever pandemic potential in Africa: a multistage analysis. Lancet 390(10113):2662–2672. https://doi.org/10.1016/S0140-6736(17)32092-5
Plowright RK, Eby P, Hudson PJ, Smith IL, Westcott D, Bryden WL et al (2015) Ecological dynamics of emerging bat virus spillover. Proc Biol Sci 282(1798):20142124. https://doi.org/10.1098/rspb.2014.2124
Plowright RK, Parrish CR, McCallum H, Hudson PJ, Ko AI, Graham AL, Lloyd-Smith JO (2017) Pathways to zoonotic spillover. Nat Rev Microbiol 15:502–510. https://doi.org/10.1038/nrmicro.2017.45
Poletti P, Visintainer R, Lepri B, Merler S (n.d.) The interplay between individual social behavior and clinical symptoms in small clustered groups. BMC Infect Dis 17(1):521. https://doi.org/10.1186/s12879-017-2623-2
Rao SR, Salins N, Joshi U, Patel J, Remawi BN, Simha S et al (2022) Palliative and end-of-life care in intensive care units in low- and middle-income countries: a systematically constructed scoping review. J Crit Care 71:154115. https://doi.org/10.1016/J.JCRC.2022.154115
Rathish, B., & Vaishnani, K. (2019). Nipah Virus-StatPearls-NCBI bookshelf nipah virus pathophysiology treatment/management enhancing healthcare team outcomes. 2–5
Reny TT, Barreto MA (2022) Xenophobia in the time of pandemic: othering, anti-Asian attitudes, and COVID-19. Politics, Groups, and Identities 10(2):209–232. https://doi.org/10.1080/21565503.2020.1769693
Rosa WE, Ahmed E, Chaila MJ, Chansa A, Cordoba MA, Dowla R et al (2022) Can you hear us now? Equity in global advocacy for palliative care. J Pain Symptom Manag 64(4):e217–e226. https://doi.org/10.1016/j.jpainsymman.2022.07.004
Sardar S, Abdul-Khaliq I, Ingar A, Amaidia H, Mansour N (2020) "COVID-19 lockdown: A protective measure or exacerbator of health inequalities? A comparison between the United Kingdom and India." A commentary on 'the socio-economic implications of the coronavirus and COVID-19 pandemic: a review'. Int J Surg. *(London, England)* 83:189–191. https://doi.org/10.1016/J.IJSU.2020.09.044
Sarfraz Z, Sarfraz A, Jaiswal V, Poudel S, Bano S, Hanif M et al (2022) The past, present and future of COVID-19 associated Mucormycosis: a rapid review. J Prim Care Community Health 13:21501319221099476. https://doi.org/10.1177/21501319221099476
Sas MA, Mertens M, Isselmou E, Reimer N, El Mamy BO, Doumbia B, Groschup MH (2017) Crimean-Congo hemorrhagic fever virus-specific antibody detection in cattle in Mauritania. Vector Borne Zoonotic Dis 17(8):582–587. https://doi.org/10.1089/vbz.2016.2084
Sayed A, Bottu A, Qaisar M, Mane MP, Acharya Y (2019) Nipah virus: a narrative review of viral characteristics and epidemiological determinants.

Public Health 173:97–104. https://doi.org/10.1016/j.puhe.2019.05.019
Sbarra AN, Rolfe S, Nguyen JQ, Earl L, Galles NC, Marks A et al (2021) Mapping routine measles vaccination in low- and middle-income countries. Nature 589(7842):415–419. https://doi.org/10.1038/s41586-020-03043-4
Sedaghat MM, Sarani M, Chinikar S, Telmadarraiy Z, Moghaddam AS, Azam K et al (2017) Vector prevalence and detection of Crimean-Congo haemorrhagic fever virus in Golestan Province, Iran. J Vector Borne Dis 54(4):353. https://doi.org/10.4103/0972-9062.225841
Seimenis A (2012) Zoonoses and poverty - a long road to the alleviation of suffering. Vet Ital 48(1):5–13
Selarka L, Sharma S, Saini D, Sharma S, Batra A, Waghmare VT et al (2021) Mucormycosis and COVID-19: an epidemic within a pandemic in India. Mycoses 64(10):1253–1260. https://doi.org/10.1111/myc.13353
Shahbazi N, Firouz SK, Karimi M, Mostafavi E (2019) Seroepidemiological survey of Crimean-Congo haemorrhagic fever among high-risk groups in the west of Iran. J Vector Borne Dis 56(2):174–177. https://doi.org/10.4103/0972-9062.263720
Shannon G, Morgan R, Zeinali Z, Brady L, Thereza Couto M, Devakumar D et al (2022) Racism, xenophobia, discrimination, and health 3 intersectional insights into racism and health: not just a question of identity. Lancet 400:2125–2136. https://doi.org/10.1016/S0140-6736(22)02304-2
Shapiro-Mendoza CK, Rice ME, Galang RR, Fulton AC, VanMaldeghem K, Prado MV et al (2017) Pregnancy outcomes after maternal zika virus infection during pregnancy - U.S. territories, January 1, 2016-April 25, 2017. MMWR Morb Mortal Wkly Rep 66(23):615–621. https://doi.org/10.15585/mmwr.mm6623e1
Song Y-J, Park W-J, Park B-J, Lee J-B, Park S-Y, Song C-S et al (2014) Hepatitis E virus infections in humans and animals. Clin Exp Vaccine Res 3(1):29–36. https://doi.org/10.7774/cevr.2014.3.1.29
Sorvillo TE, Rodriguez SE, Hudson P, Carey M, Rodriguez LL, Spiropoulou CF et al (2020) Towards a sustainable one health approach to Crimean-Congo hemorrhagic fever prevention: focus areas and gaps in knowledge. Trop Med Infect Dis 5(3):113. https://doi.org/10.3390/tropicalmed5030113
Spengler JR, Estrada-Peña A, Garrison AR, Schmaljohn C, Spiropoulou CF, Bergeron É, Bente DA (2016) A chronological review of experimental infection studies of the role of wild animals and livestock in the maintenance and transmission of Crimean-Congo hemorrhagic fever virus. Antivir Res 135:31–47. https://doi.org/10.1016/j.antiviral.2016.09.013
Stewart CR, Deffrasnes C, Foo CH, Bean AGD, Wang L-F (2018) A functional genomics approach to henipavirus research: the role of nuclear proteins, MicroRNAs and immune regulators in infection and disease. Curr Top Microbiol Immunol 419:191–213. https://doi.org/10.1007/82_2017_28
Stoner MCD, Pettifor A, Edwards JK, Aiello AE, Halpern CT, Julien A et al (2017) The effect of school attendance and school dropout on incident HIV and HSV-2 among young women in rural South Africa enrolled in HPTN 068. AIDS 31(15):2127–2134. https://doi.org/10.1097/QAD.0000000000001584
Surtees R, Dowall SD, Shaw A, Armstrong S, Hewson R, Carroll MW et al (2016) Heat shock protein 70 family members interact with Crimean-Congo hemorrhagic fever virus and Hazara virus nucleocapsid proteins and perform a functional role in the Nairovirus replication cycle. J Virol 90(20):9305–9316. https://doi.org/10.1128/JVI.00661-16
Taylor LH, Latham SM, Woolhouse ME (2001) Risk factors for human disease emergence. Philos Trans R Soc Lond Ser B Biol Sci 356(1411):983–989. https://doi.org/10.1098/rstb.2001.0888
Terefe M, Solomon C, Menza K, Bedemo A (2017) Determinants of urban poverty: the case of Nekemte town, Eastern Wollega zone of Oromia regional. State, Journal of Poverty Journal
Terrero-Salcedo D, Powers-Fletcher MV (2020) Updates in laboratory diagnostics for invasive fungal infections. J Clin Microbiol 58(6):1–11. https://doi.org/10.1128/JCM.01487-19
Thompson GR, Le T, Chindamporn A, Kauffman CA, Alastruey-Izquierdo A, Ampel NM et al (2021) Global guideline for the diagnosis and management of the endemic mycoses: an initiative of the European confederation of medical mycology in cooperation with the international society for human and animal mycology. Lancet Infect Dis 21:e364–e374. https://doi.org/10.1016/S1473-3099(21)00191-2
Tice, P. (2013). Substance use among 12th grade aged youths by dropout status
Tomley FM, Shirley MW (2009) Livestock infectious diseases and zoonoses. Philos Trans R Soc Lond Ser B Biol Sci 364:2637–2642. https://doi.org/10.1098/rstb.2009.0133
Tran DN, Manji I, Njuguna B, Kamano J, Laktabai J, Tonui E et al (2020) Solving the problem of access to cardiovascular medicines: revolving fund pharmacy models in rural Western Kenya. BMJ Glob Health 5(11):e003116. https://doi.org/10.1136/BMJGH-2020-003116
Trolle H, Forsberg B, King C, Akande O, Ayres S, Alfvén T, Elimian K (2023) A scoping review of facilitators and barriers influencing the implementation of surveillance and oral cholera vaccine interventions for cholera control in lower- and middle-income countries. BMC Public Health 23(1):455. https://doi.org/10.1186/S12889-023-15326-2
Undurraga EA, Carias C, Meltzer MI, Kahn EB (2017) Potential for broad-scale transmission of Ebola virus disease during the West Africa crisis: lessons for the Global Health security agenda. Infect Dis Poverty 6(1):159. https://doi.org/10.1186/s40249-017-0373-4
van Roon A, Maas M, Toale D, Tafro N, van der Giessen J (2019) Live exotic animals legally and illegally imported via the main Dutch airport and consider-

ations for public health. PLoS One 14(7):e0220122. https://doi.org/10.1371/journal.pone.0220122

Vaughan A, Aarons E, Astbury J, Balasegaram S, Beadsworth M, Beck CR et al (2018) Two cases of monkeypox imported to the United Kingdom, september 2018. Euro Surveill 23(38):1800509. https://doi.org/10.2807/1560-7917.ES.2018.23.38.1800509

Wagstaff, A., Claeson, M., Hecht, R. M., & Gottret, P. (2003). Chapter 9. Millennium development goals for health : what will it take to accelerate progress ? health (San Francisco), 181–194.. https://doi.org/NBK11716 [bookaccession]

Wang C, Tee M, Roy AE, Fardin MA, Srichokchatchawan W, Habib HA et al (2021) The impact of COVID-19 pandemic on physical and mental health of Asians: A study of seven middle-income countries in Asia. PLoS One 16(2):e0246824. https://doi.org/10.1371/journal.pone.0246824

Wasfi F, Dowall S, Ghabbari T, Bosworth A, Chakroun M, Varghese A et al (2016) Sero-epidemiological survey of Crimean-Congo hemorrhagic fever virus in Tunisia. Parasite 23:10. https://doi.org/10.1051/parasite/2016010

Williams SC, Van Oosterwijk JG, Linske MA, Zatechka S, Richer LM, Przybyszewski C et al (2020) Administration of an orally delivered substrate targeting a mammalian zoonotic pathogen reservoir population: novel application and biomarker analysis. Vector Borne Zoonotic Dis. Https://Home.Liebertpub.Com/Vbz 20(8):603–612. https://doi.org/10.1089/VBZ.2019.2612

Wolfe ND, Dunavan CP, Diamond J (2007) Origins of major human infectious diseases. Nature 447(7142):279–283. https://doi.org/10.1038/nature05775

Woodward A, Sheahan K, Martineau T, Sondorp E (2017) Health systems research in fragile and conflict affected states: a qualitative study of associated challenges. Health Res Policy Syst 15(1):1–12. https://doi.org/10.1186/s12961-017-0204-x

Woolhouse MEJ, Gowtage-Sequeria S (2005) Host range and emerging and reemerging pathogens. Emerg Infect Dis 11(12):1842–1847. https://doi.org/10.3201/eid1112.050997

World Health Organisation(WHO) (2016) Neglected tropical diseases. https://doi.org/10.1037/0021-9010.93.1.170

World Health Organization (2020) Prioritizing diseases for research and development in emergency contexts. *World-Health-Organization*. https://www.who.int/activities/prioritizing-diseases-for-research-and-development-in-emergency-contexts

World Health Organization. (2022). WHO fungal priority pathogens list to guide research, development and public health action licence: CC BY-NC-SA 30 IGO, Vol. 1, pp. 1–48

World Health Organization (WHO) (2019) A tripartite guide to addressing zoonotic diseases in countries. World Organisation for Animal Health (OIE). https://books.google.com.mt/books/about/Taking_a_Multisectoral_One_Health_Approa.html?id=uDC1DwAAQBAJ&printsec=frontcover&source=kp_read_button&redir_esc=y#v=onepage&q&f=false

Yonzan N, Cojocaru A, Lakner C, Greszon M, Narayan A (2022) The impact of COVID-19 on poverty and inequality: evidence from phone surveys. World Bank blogs. https://blogs.worldbank.org/opendata/impact-covid-19-poverty-and-inequality-evidence-phone-surveys. Accessed 20 Apr 2023

Zakham F, Alaloui A, Levanov L, Vapalahti O (2019) Viral haemorrhagic fevers in the Middle East. Rev Sci Tech 38(1):185–198. https://doi.org/10.20506/rst.38.1.2952

Zere E, Moeti M, Kirigia J, Mwase T, Kataika E (2007) Equity in health and healthcare in Malawi: analysis of trends. BMC Public Health 7(1):78. https://doi.org/10.1186/1471-2458-7-78

Zinsstag J, Utzinger J, Probst-Hensch N, Shan L, Zhou X-N (2020) Towards integrated surveillance-response systems for the prevention of future pandemics. Infect Dis Poverty 9(1):140. https://doi.org/10.1186/s40249-020-00757-5

4 Preparedness and Response: Outlook Post COVID-19 Pandemic and SDG3d

Abstract

The COVID-19 pandemic has affected populations globally, from loss of lives, livelihoods and socially through the sudden changes in all aspects of life. While COVID-19 variants are being recorded in many places globally, vaccination rates skewed with LMIC struggling with lower vaccination rates and access how can response to emerging and re-emerging disease be sustainable? How can LMIC achieve sustainable management of health emergencies? With Avian influenza, MERS, Ebola, Marburg, Zika and many emerging diseases in pockets of the world and LMIC mostly affected, how can we ensure that livelihoods of these populations are sustained during outbreak response and management? The events during the COVID-19 pandemic and the current Monkeypox outbreak have shown a growing divide in global health response between HIC and LMIC. Unlike smallpox which was eradicated with joint efforts to end the dreadful disease, the current efforts display a rather grim picture. Sustainable response thus requires a holistic approach that takes into consideration all manner of disease, population, geographic setting, social, economic, and environmental impacts. The COVID-19 pandemic has revealed gaps in health systems, governance, regional and global collaboration, access, and various inequalities that exist. A sustainable response would require therefore addressing these gaps and working towards attaining SDG3d in LMIC.

Keywords

Sustainable development goals (SDG) · Outbreak preparedness · Response · SDG3d · Post-Covid-19 pandemic · Health emergencies · Low and Middle Income Countries (LMIC) · Monkeypox · Ebola

The COVID-19 pandemic has affected populations globally, from loss of lives, livelihoods and socially through the sudden changes in all aspects of life. While COVID-19 variants are being recorded in many places globally, vaccination rates skewed with LMIC struggling with lower vaccination rates and access how can response to emerging and re-emerging disease be sustainable? How can LMIC achieve sustainable management of health emergencies? With Avian influenza, MERS, Ebola, Marburg, Zika and many emerging diseases in pockets of the world and LMIC mostly affected, how can we ensure that livelihoods of these populations are sustained during outbreak response and management?

A rather intriguing pattern of response to infectious disease has been observed in the past decades

F. A. Mphande, *Sustainable Health in Low and Middle Income Countries*, Sustainable Development Goals Series, https://doi.org/10.1007/978-981-99-4254-1_4

with some of the major outbreaks of Ebola, COVID-19 and the recent Monkeypox virus. While outbreaks maybe ongoing in LMIC if they remain in those regions, it may not catch as much media attention until it becomes a threat to the global North. This type of response has resulted in limited response when it comes to resources, including investment in vaccines, and treatment in the affected populations. The response to COVID-19 pandemic and the Monkeypox outbreak in Europe is evidence to such a response. While various treatments and vaccines were made available in high income communities, LMIC remain behind on accessing such crucial resources. Does a disease have to affect populations countries in the global North to solicit certain types of response? SDG3 clearly point to “Ensure healthy lives and promote well-being for all at all ages” a point which has been clearly missed in the current response efforts to COVID-19 and monkeypox disease outbreak. The events during the COVID-19 pandemic and the monkeypox outbreak have shown a growing divide in global health response between HIC and LMIC. Unlike smallpox which was eradicated with joint efforts to end the dreadful disease, the current efforts display a rather grim picture. With response moving more towards restricting movement of certain populations rather than providing equitable access to treatment and preventive care. It is worth revisiting the efforts made during the global eradication of smallpox for key lessons to attain SDG3.3 “Fight communicable diseases.”

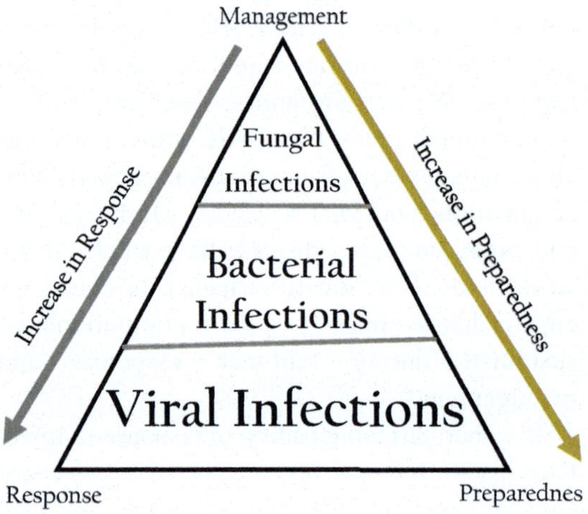

4.1 Global Response Vs Local Response

A number of confirmed cases of monkeypox also known as mpox, disease with no links of travel to endemic countries was reported in 2022, with a notification alert initiated from the United Kingdom (UK Government 2022).

According to the European Centres for Disease Control (ECDC), 21 127 confirmed cases of monkeypox have been reported from 29 EU/EEA countries since the beginning of the outbreak in 2022. On a global scale 84,916 confirmed cases and 1355 probable cases have been reported from 110 countries from six WHO regions. Apart from West African countries where the disease is endemic, the remaining affected countries have reported the disease to occur in men who have sex with men (European Centre for Disease Prevention and Control 2022; WHO 2022a). By December 2022, WHO region of the Americas had the highest number of cases reported (WHO 2022a).

The mechanism of transmission of monkeypox is still understudy although prolonged contact during sex has been reported as one of the possible mechanisms in the current outbreak (Rodríguez et al. 2022; UK Government 2022). Historically monkeypox was reported as a zoonotic disease endemic to Central Africa region in areas with close proximity to the rain forest (“Monkeypox,” n.d.). Until 2022 reservoirs of the disease have included rope squirrels, tree squirrels, Gambian pouched rats, dormice, non-human primates, and other species although further studies are required to understand the reservoirs of the disease within the community. The current outbreak in Europe has shown a different pattern of disease with no link to animal reservoirs (yet) but human -to- human transmission with high incidence in men who have sex with men (MSM); although a few cases unrelated to sex have also been reported. Further studies are required to understand the changes in the virus behaviour, possible new reservoirs as well as changes in human behaviour. A study in Spain reported 85.8% cases had reported intimate and prolonged contact during sex, while only less than 15% of the cases were reported from close contacts unrelated to sex (Rodríguez et al. 2022).

Various countries managed the monkeypox outbreak following WHO recommendations as well as regional and national health policies governing disease prevention and control. Some of the mechanisms of response that were used are outlined (Table 4.1).

4.1.1 Preparedness and Response Outbreaks and Health Emergencies

Governance is an important aspect in management and coordination of an infectious disease outbreak and or any health emergency. Issuing alerts in a timely manner ensures that populations are well prepared and aware of the current updates and allows for rapid and coordinated response between various stakeholders (Table 4.1).

Preparing protocols and ensuring that these protocols are available for use at the point of care and or any places that there are routines that require strict protocols. For example in case of an outbreak having clear protocols including those for early detection, case finding, contact tracing, case definitions could ensure better response and management of the situation. Existence and implementation of such protocols makes a difference between attaining SDG3d in LMIC. Unfortunately not all countries have achieved this. SDG3d focuses on emergency preparedness and strengthening the capacity of all countries in particular developing countries for early warning, risk reduction and management of national and global health risks. Different levels of preparedness exist in individual countries in LMIC. Stakeholders should be able to know how to manage a case from detection, treatment, contact tracing and management of the cases after recovery within their setting. Response does not end at recovery of the case but ensuring that there is no further spread of the disease from the reservoir host as well as the recovered patients, hence need for sustained screening for survivors of disease such as Ebola, COVID-19 and other priority diseases (Balachandar et al. 2020; Diallo et al. 2016, 2019; Queeneth Ojoma et al. 2021; Subissi et al. n.d.; Xu et al. 2020) (Table 4.1).

Table 4.1 Preparedness and response strategies (Rodríguez et al. 2022)

Preparedness	Response
Governance	Issue alerts to all stakeholders to pursue a rapid and coordinated response
Prepare protocols	A national protocol for early detection and case and contact management was approved and made available by the national alert board and coordinated by the ministry of health 3 days after detection of suspected cases
Risk assessments	Rapid risk assessment and reporting ensure that risks are identified, and risk management measure are implemented as early and rapid as possible
Reporting	Situation reports that are updated regularly to ensure that various stakeholders and communities are well informed
Communication and consultation	Communication and consultation with relevant stakeholders is essential in timely sharing of information ensuring early and effective preparedness
Stakeholder partnership	Stakeholder partnerships are crucial for management and implementation of preparedness and response strategies. Some of the factors involved with stakeholder partnerships include information dissemination, development of protocols and policies, as well as implementation of intervention programmes

Rapid and thorough risk assessment ensures that all aspects of disease spread are explored, and control measures are in place to abate as well as manage the risks. These include the distribution of the disease within a population, source of the disease, possible host (or hosts), susceptible populations, risk populations, transmission patterns as well as protective factors. Knowledge of these factors is possible through data collection using a comprehensive surveillance system able to collect as much information as possible. A well planned and sustained surveillance system provides data that can assist in providing disease patterns and information that can be used to provide early warning systems and manage risks (Brierley et al. 2016; Galan et al. 2016; Jones et al. 2008; Li et al. 2019; Murray et al. 2015; Peel et al. 2019; Van Kerkhove et al. 2015). The data that can be obtained through surveillance include detection of risk pathogens, risk populations, provide distribution patterns and further information on the possible reservoir and host. Communicating the risks and management protocols through situation reports ensures that communities and various stakeholders are well informed on the situation and are able to take the necessary measures to reduce the risk of spread of the disease (Table 4.1).

Communication is key in achieving effective preparedness and response. Communication with various stakeholders requires proper planning ensuring that the right modes of communication are used, and that the information being disseminated is clearly understood by the stakeholders. Ensuring that the right information is communicated will reduce the risk of misunderstanding which could hinder effective preparedness and response(Table 4.1).

On the African context, various changes were suggested to improve preparedness and response efforts. These were made in response to that gaps that were observed during the COVID-19 preparedness and response efforts. Five key improvements were identified to better health systems. These include,

1. *Strengthening public health capabilities*: There is need for strengthening of public health facilities at national and regional level. At national level this will be achieved through improved infrastructure and enhancing capacities including diagnostics, therapeutics, research and development, and surveillance. The national public health institutes will work as drivers for implementing International Health Regulations (IHR) while regional networks will be used to enhance capacities such as surveillance, laboratory, emergency operation centres, and innovative information systems (Nkengasong et al. 2017).
2. *Incentives, introducing a system of rewards and penalties*: A system of reward and penalties for those who have achieved or failed to achieve the set capacities could be used as an incentive for countries to meet their required IHR targets. In this system, each country will be given an aggregate score through a joint external evaluation which will give an aggregate score depending on the capacities achieved. The mechanism how this type of system could be implemented is yet to be established (Nkengasong et al. 2017).
3. *Community participation*: Communities are an integral part of preparedness and response. Willingness and acceptance of interventions by the community will highly likely affect the communities adherence and further affect the outcomes of the implementation of the programs. Encouraging community participation and engagement and decentralization of IHR implementation could assist in better preparedness and response both at national and regional levels (Nkengasong et al. 2017).
4. *Enhancing Capacity for public health workers:* The capacity of public health workers in Africa should be enabled to meet IHR needs at national level. This includes adapting and recognizing community healthcare worker programmes, as well as strengthening field epidemiology training programmes. There is need to enhance capacities of sub-specialties such as field laboratory leadership programmes, public health informatics training programmes, and through public health management of hazards and pandemics (Nkengasong et al. 2017).
5. *Data management and Sharing:* Challenges in governance, sharing and management of public health data are not new on the African continent. Increasing the continents capabilities in generating quality public health data in real time is something to strive towards. Sharing data through regional networks such as the recently established CDC's Regional Integrated Surveillance and Laboratory Networks is essential. This will reduce costs and prevent duplication of data as well as research efforts to generate already existing

data. Data sharing will also enable timely alerts to neighbouring countries which would benefit in preparedness and response efforts to disease threats on the continent and beyond (Nkengasong et al. 2017).

The five improvements are in line with SDG3d focusing on strengthening capacities and capabilities that will assist in early detection and management of health risks and emergencies on the African continent.

4.2 Tools for Preparedness and Response

What are the tools that are required for preparedness and response for health emergencies of varying magnitudes?

4.2.1 Prepare for Emerging Pathogens with Varying Disease Patterns and Characteristics

The COVID-19 pandemic laid bare the insufficiencies in preparedness and response in both HIC and LMIC. In general deficiencies were observed in the limited policies and strategies in place to tackle health emergencies such as the COVID-19 pandemic. The highly infectious nature of the disease, the transmission patterns, and gaps in knowledge in the epidemiology of the virus contributed to the complexity of the disease. While virus spillovers have occurred before in highly infectious diseases such as Ebola, Marburg and Nipa virus disease, the self-limiting nature of human-to-human transmission of some of these diseases have allowed for limited spread of the disease through containment. On the contrary, COVID-19 sustained human -to - human transmission with high rate of mutations led to emergence of variants some of which were highly transmissible and virulent. These characteristics of COVID-19 were not anticipated and not prepared for. Despite the rapid development and dissemination of vaccines, the increased number of mutations and emerging variants has complicated these efforts (El-Shabasy et al. 2022; Fernandes et al. 2022; Haque and Pant 2022). The dynamic nature of COVID-19 presented varying challenges for the health system, public health personnel, and existing infrastructure including policies and strategies on infectious disease control and prevention (El-Shabasy et al. 2022; Fernandes et al. 2022; Haque and Pant 2022).

Studies assessing the management of the COVID-19 pandemic identified critical challenges for the management not only in the health systems but also in the policies, strategies as well as existing infrastructure (Khankeh et al. 2021; Shirani et al. 2020). While limited evidence, scientific controversies, poor social prevention and social inequalities, could have affected response, burnout and sustained workload among healthcare workers was also a challenge (Tusabe et al. 2022). Improper management of resources and equipment, the lack of a guidelines for contact tracing, and patient flow management had an effect on management of cases. Socially, mental health problems in the community were another challenge that was identified (Khankeh et al. 2021). It is therefore crucial to critically analyze the achievements and challenges countries faced during the COVID-19 pandemic. The outcomes would assist in setting up early warning systems as well as improve risk management of health emergencies in LMIC.

4.2.2 Prepare for Social and Economic Impacts of the Disease on the Society

Social and economic impacts of COVID-19 including management of vaccination programmes as well as existing disease prevention campaigns such as HIV and TB control were experienced globally (Harris et al. 2021; Hussain et al. 2020; Iglesias Martínez et al. 2022; Izudi et al. 2022; Klinton et al. 2021; Madzima et al. 2022; Turk and Mitra 2021). Mental health took its toll on populations as they tried to navigate the various prevention measures including, lockdown, travel restrictions, coping with loss of family members and long term effects of the disease for those who survived the disease as well as caregivers. A report from OECD outlined insufficient preparedness in the area of human and financial costs (McClarty et al. 2022; Pillay et al. 2021; Xie et al. 2022). Although governments managed to take swift and massive action to mitigate the economic and financial effects of the pandemic there is need for careful monitoring for longer-term budgetary costs of these measures. Transparency was another area that was identified. There was a need for frequent targeted crisis communication with stakeholders and the public in risk-related decision-making. Gaps in the evidence base were acknowledged, evidence is crucial for better preparedness. Insufficient evidence could hinder effective response. There is need for further assessment of the effectiveness of lockdowns, which had varying health, social and economic impacts on populations worldwide. Impact of response measures including lockdown on social issues such as domestic violence, alcohol consumption, youth, and mental health, have to be critically considered (OECD 2022).

4.3 Case Studies in Preparedness and Response Lessons from COVID-19 and Monkeypox Disease

A monkeypox outbreak outside endemic areas has been reported since 2022, the first of its kind with no links of travel to endemic areas, this outbreak has caught the global attention. The outbreak was one of the major outbreaks of an infectious diseases crossing continents after the COVID-19 pandemic. As such it is of interest to see how countries were prepared and responded to the disease. The case studies below outline some of the plans and techniques taken by countries to combat the disease.

4.3.1 Case 1: Epidemiologic Features and Control Measures During Monkeypox Outbreak, Spain, June 2022 (Rodríguez et al. 2022)

The case definition for monkeypox in Spain was as follows:

Confirmed case-patient as a person who had a PCR identification of the monkeypox genome or who had a positive result in a generic PCR for Orthopoxvirus in a clinical sample.

4.3.1.1 Response Strategy

Following the procedures of the National Early Warning and Rapid Response System,

1. All key stakeholders were alerted to pursue a rapid and coordinated response
2. A national protocol for early detection and case and contact management was approved and made available by the National Alert Board and coordinated by the Ministry of Health 3 days after detection of suspected cases.
3. A rapid risk assessment for Spain was conducted and reported (9), and situation reports were updated regularly (3)
4. Early consultation and exchange with relevant scientific societies led to publication of an atlas that contained differential diagnoses for monkeypox skin lesions.
5. Created partnerships with relevant stakeholders, including with the lesbian, gay, transgender, bisexual, intersex, and queer (LGTBIQ) community was seen as pivotal, and the Ministry of Health involved its Advisory and Counselling Board of nongovernmental organizations in the response to promote the engagement of the LGTBIQ community.
6. Dissemination of key health messages which were developed, and were made publicly available
7. Building on previous health campaign experience following the general principles of the World Health Organization and European Centre for Disease Prevention and Control

Identifying and tracking contacts was one of the challenges described in this paper. Cases/Patients could; be hesitant to provide the identities of their contacts; or were not able to do so as exposure occurred anonymously with previously unknown persons. In other circumstances it was difficult to ascertain the exact dates on which transmission might had occurred. Despite these challenges, strategies consistent with SDG3d are in play including early detection and risk management.

4.3.2 Case 2: A New Public Health Order for Africa's Health Security (Nkengasong et al. 2017)

The paper outlined five key improvements that could usher change in the health security of the African continent. The five key improvements are necessary.

1. *Strengthening public health capabilities*, with national public health institutes as the drivers of IHR implementation. Where there should be improved infrastructure and enhanced capacities for integrated national and regional networks for disease surveillance, including laboratories, emergency operation centres, and innovative information systems.
2. *Implementation of a reward and penalty system* along with financial incentives to acknowledge whether progress has been made in implementing the core capacities of IHR through rigorous objective external stepwise assessments, such as the joint external evaluation (JEE), creating an IHR aggregate score for each country.
3. Decentralised implementation of the IHR to the subnational levels of the health service, with a strong community engagement.
4. Enabling the African public health workforce to meet the IHR needs and other commitments at a national level, including by adapting and recognising community health-care worker programmes, field epidemiology training programmes with subspecialties to meet IHR requirements, field laboratory leadership programmes, public health informatics training programmes, and through public health management of hazards and pandemics.
5. Establishing governance procedures for the management of public health data sharing. A culture of data sharing must be promoted in the new public health order to combat the current data sharing challenges on the continent. Sharing disease data in a timely fashion through recently established public health constructs such as the Africa CDC's Regional Integrated Surveillance and Laboratory Networks will be critical to advance the implementation of IHR and enhance African health security. Not sharing disease data for public health is costly to countries as it can lead to duplicate research efforts to generate data that might already exist and might have enabled a neighbouring country to be better alerted, prepared, and able to respond to a disease threat.

This paper describes plans on how SDG3d could be achieved on the African continent and its respective countries. Early detection systems are still in their infancy and in some cases not yet in place.

4.3.3 Case 3: Outbreak of Human Monkeypox in Nigeria in 2017–2018: A Clinical and Epidemiological Report (Yinka-Ogunleye et al. 2019)

In September, 2017, human monkeypox reemerged in Nigeria, 39 years after the last reported case. The study aimed to describe the clinical and epidemiological features of the 2017–2018 human monkeypox outbreak in Nigeria.

One hundred and twenty-two confirmed or probable cases including seven deaths (case fatality rate 6%) of human monkeypox were recorded in 17 states. The age range of people infected with monkeypox was 2 days to 50 years (median 29 years [IQR 14]). Sixty nine percent (84) were male. The distribution of cases and contacts suggested both primary zoonotic and secondary human-to-human transmission. Two cases of health-care-associated infection were recorded. Genomic analysis suggested multiple introductions of the virus and a single introduction along with human-to-human transmission in a prison facility.

4.3.3.1 Response Strategy

1. The Nigeria Centre for Disease Control activated a national outbreak response, including enhanced monkeypox surveillance.12
2. After detection of the first suspected case, the Nigeria Centre for Disease Control developed standard case definitions (panel) and interim guidelines based on previous outbreak reports from DR Congo and the USA.
3. Case-based surveillance was instituted, and an electronic Surveillance Outbreak Response and Analysis System introduced to improve digitalisation and timeliness of the surveillance system.
4. A detailed and standardised case investigation form was developed on the basis of available literature and guidelines for human monkeypox. The form was used by clinicians, who clinically examined patients, and surveillance officers (public health officers employed by local and state government to work with the community and health facilities to detect and report priority diseases) and epidemiologists, who reviewed patients' clinical records to collect clinical and epidemiological data.
5. Data collected included sociodemographic characteristics, clinical symptoms and signs, smallpox vaccination history, detailed contact history, and other surveillance-related information. In many cases, clinical photographs of rashes and other lesions were also documented (after obtaining patients' verbal consent).
6. All data were collected for the purpose of diagnosis and outbreak control as part of constitutional disease control procedures in Nigeria. Thus, specific ethics approval was not required.
7. Laboratory investigations: At least one type of specimen (either blood, lesion swab, or crust i.e., dried rash debris—serum, pus, or blood—on the skin surface) was collected from each patient for investigation. Real-time PCR, serology, and culture were done on the samples. In the first 6 weeks of the outbreak, PCR testing was done at Institut Pasteur de Dakar (Dakar, Senegal).

- *Case definitions used during the 2017–2018 Nigeria human monkeypox outbreak*
- *Suspected case:* any person presenting with a history of sudden onset of fever, followed by a vesiculopustular rash occurring mostly on the face, palms, and soles of feet
- *Confirmed case*: any suspected case with laboratory confirmation (ie, viral identification by real-time PCR, antibody detection, or viral isolation)
- *Probable case:* any suspected case in whom laboratory testing could not be done but who could be epidemiologically linked with a confirmed case
- *Case:* any probable or confirmed case
- *Contact person:* any person who has no symptoms but had been in physical contact with a suspected case or with body fluids (i.e., skin secretions, oral secretions, urine, faeces, vomit, or blood) of a case in the past 3 weeks
- *Monkeypox death:* any death in a confirmed monkeypox case that occurred during the course of monkeypox rash illness and that has no other suggestive cause of death

The paper describes efforts in surveillance, case detection, and diagnosis on monkeypox cases. From the narrative there are gaps to achieve SDG3d. International collaboration for example working with laboratories from neighboring countries are commendable; considerations should be taken on costs, logistics and time constraints.

4.3.4 Case 4: Bhutan's Preparedness for Monkeypox Outbreak (Tamang and Dorji 2022)

Background and Setting: Bhutan provides free medical services to its entire population. There are no private hospitals in the country although a few private "diagnostic centres" provide basic testing facilities. The country has state-run hospitals are categorized at three tiers: primary (20-bedded hospitals, primary health centres), secondary (district & general hospitals) and tertiary (two regional referral hospitals and a national referral hospital). As of January 2022, the country had 279 doctors, 1608 nurses and 2285 other health practitioners (pharmacists, dental surgeons, physiotherapists, laboratory personnel, imaging technicians, etc).6

4.3.4.1 Monkey Pox Preparedness and Response

With the ongoing monkeypox outbreak, Bhutan's Ministry of Health announced a series of measures to prevent, or respond rapidly to, an outbreak of monkeypox. The following were the measures:

- *Mandatory declaration forms*: The Ministry implemented mandatory declaration forms for all incoming travellers returning from affected countries.
- *Clinical management guidelines*: The government released its national clinical management guidelines for monkeypox in June 2022.
- *Listed Monkeypox as a notifiable disease*: Monkeypox is now a notifiable disease to be reported through its existing disease surveillance system—National Early Warning, Alert Response Surveillance (NEWARS).
- *Submission of real-time/periodic reports*: Through the NEWARS portal, all health facilities in the country submit real-time and/or periodic reports to the Royal Centre for Disease Control (RCDC) in Thimphu.
- *Changes and upgrades to the health system*: Several upgrades and changes initiated to the health system in response to COVID-19 could aid in responding to an outbreak of monkeypox.
- *Institution of a task force*: The national, regional, and local task forces instituted during COVID-19 could be swiftly reactivated.
- *Availability of PPE and use*: Any need for personal protective equipment (PPE) could be addressed rapidly: the Ministry of Health has gained experience in arranging supply and has leftover stock of PPE, and all health personnel are familiar with its use.
- *Upgrade of health infrastructure*: Critical health infrastructures were upgraded: new makeshift hospitals and additional ICU beds were set up across the country and existing hospitals upgraded with new equipment.

The stigma associated with monkeypox due to the reportedly higher incidence among men who have sex with men (MSM) and/or sexual transmission could deter health seeking behaviour and delay detection. Some challenges that were experienced in Bhutan include: fatigue to public health emergencies, a vulnerable economy, and lack of experience and knowledge of health professionals in managing monkeypox including limited diagnostic capacity.

The prompt preparedness and response described in this paper, is an example of a health system capable of making rapid changes in an emergency situation. Challenges at community level (stigma), national level (economy) and healthcare system (health professional and infrastructure) were also reported. Despite these challenges steps towards SDG3d are promising.

4.3.5 Case 5: Latin America: Situation and Preparedness Facing the Multi-Country Human Monkeypox Outbreak (Rodriguez-Morales et al. 2022)

Background and Setting: Multiple concerns have been raised, mainly from the healthcare sector, regarding currently available treatments and vaccination. Despite the absence of specific therapeutic alternatives for monkeypox, drugs with proven experimental efficacy and potential clinical impact such as cidofovir (especially its lipid conjugate brincidofovir) and tecovirimat, are not widely available in the region. Also, although monkeypox vaccination has been implemented for contacts of positive cases, at this stage, neither non-replicating/replicating-deficient live vaccinia virus-based vaccines with low reactogenicity, such as JYNNEOS®, nor classical anti-smallpox vaccines are available in most Latin American countries.

4.3.5.1 Preparedness and Response

- *Enhanced molecular testing capacity*: Due to the COVID-19 pandemic, several countries in the region have ramped up their current molecular testing capacity and have established
- *Data sharing through networks*: broad laboratory networks sharing genomic surveillance data, resulting in better preparedness against other emerging threats such as the current monkeypox multi-country outbreak.
- *Multi sectoral approach*: Improvement in data integration between different sectors in the society, including healthcare and public health authorities,
- *Enhanced infrastructure*: enhanced sanitary infrastructure,
- *Drug efficacy and safety*: use of drugs with proven efficacy and safety
- *Availability of guidelines*: the issuing of evidence-based guidelines in multiple countries has prevented a larger-scale expansion of the disease.
- *Positive impact of COVID-19 on quality of primary care*: COVID-19 pandemic have positively impacted the quality of primary care interventions build-up of intensive care units (ICU) capacities, equipment provisions and personnel training, among other advances.3

4.3.5.2 Challenges

- *Failures in risk communication*. Strengthening epidemiological surveillance systems, and disseminating adequate information through reliable channels (official social media and web pages) with clear and assertive messages could contribute to gaining greater confidence from the broad public and assisting in the early case detection, thus halting transmission chains and preventing further outbreaks.
- *Misinformation and disinformation*: Public health professionals, physician communities and organizations, healthcare authorities and scientific experts should combat misinformation and disinformation proactively based on clear, direct, culturally responsive messaging that is free of unnecessary scientific jargon.
- *Syndemics:* the Latin American region still faces a complex scenario with multiple unfolding syndemics', including communicable diseases, such as HIV infection, malaria, tuberculosis, orthohantavirus, arboviral diseases (particularly dengue, Zika, chikungunya, and yellow fever), among other endemic diseases.
- *Poverty:* The still-prevalent large pockets of poverty present in the region are part of an unavoidable context that influences disease emergence having a higher impact in Latin America when compared to other high-income countries.
- *Re-emergence of vaccine-preventable diseases*: The Latin American region has also witnessed the re-emergence of some vaccine-preventable diseases, such as chickenpox as such, vaccination programs need to be enhanced in order to recover optimal coverage.
- *Education and training for health care workers*: There is an urgent need for healthcare workers' education on the clinical and epidemiological aspects of monkeypox, including characterization and inclusion as a part of the differential diagnoses with other diseases endemic in the Latin American region that may overlap with similar clinical findings.
- Need to train healthcare workers in co-infections including other sexually transmitted pathogens such as *Treponema pallidum*, endemic trepanomatosis like yaws (*T. palli-*

dum subsp. pertenue) and *pinta* (*T. carateum*) which are also prevalent in Latin America.

Prompt prioritization and allocation of necessary resources, strengthening epidemiological surveillance systems, and increased capacity to detect imported cases and limit onward transmission (including autochthonous cases).

The situation in Latin America as described in this paper shows some level of preparedness but there is need for strengthening of existing systems including surveillance, diagnostics, as well as management of existing vaccine preventable diseases and syndemics. The paper also reports on challenges of misinformation which is a great risk to management of disease outbreaks. Thus although strides have been made towards achieving SDG3d, more work is needed to Strengthen the existing capacities.

4.3.6 Case 6: Vaccination for Monkeypox Prevention in Persons with High-Risk Sexual Behaviours to Control On-Going Outbreak of Monkeypox Virus Clade 3 (Petersen et al. 2022)

Background: To distinguish between the various strains of Monkeypox in circulation, a new nomenclature has been proposed to distinguish recurring local cases in known enzootic regions from the current outbreak. The previous Congo Basin lineage has been classified as clade 1, the West African local cases (and incidental travellers) as clade 2, and the current outbreak outside of the African region as clade 3 (Happi et al. 2022). A recent estimate of the R0 of the ongoing outbreak of monkeypox virus clade 3 suggests it may be substantially higher than 1, thus sustaining an expanding outbreak in this high-risk population of young men who are too young to have had a smallpox vaccine in their childhood (Endo et al. 2022).

4.3.6.1 Preparedness and Response Strategies:

- *Protecting frontline workers and stopping the spread of the disease*: The priority should be on stopping further spread and protecting frontline health-care workers.
- *Strengthening surveillance*: If surveillance was strengthened and response timely the outbreak could be stopped at the source (Zumla et al. 2022).
- *Predisposing risk factors*: Some published data suggests that it is a particular subset of the MSM population that is at risk in this current monkeypox outbreak outside endemic regions. Perez Duque et al., reported that fifty percent of cases in early reports from Portugal and Italy were HIV-positive (Ferraro et al. 2022; Perez Duque et al. 2022). Another suggested predisposing factor is young men who are too young to have had a smallpox vaccine in their childhood (Endo et al. 2022).
- *Post Exposure Prophylaxis (PrEP)*: WHO proposes post-exposure prophylaxis (PEP) in a form of vaccine offered to contacts of cases within 4 days of first exposure, and recommended for health workers at risk, laboratory personnel working with orthopoxviruses, clinical laboratory staff performing diagnostic testing for monkeypox, and others who may be at risk as per national policy (WHO 2022a, b).
- *Use of Smallpox vaccine as PrEP:* The use of smallpox vaccine as PrEP in MSM at high risk of monkeypox virus exposure may also reduce risk of transmission into the general population.
- *Vaccine demand, roll out and equity of distribution:* The scale of PrEP could be decided in reference to the exposure risk level in each respective jurisdiction. This will facilitate a controlled roll-out of vaccines and should be implemented when vaccine production has been ramped up to meet demand. No doubt vaccine production must be increased to ensure access and equity also outside Europe and North America.

This paper links surveillance to outbreak outcomes and timely response. The lack of vaccines, therapies and the need to protect frontline workers is also emphasized. Thus, surveillance remains crucial to achieving SDG3d. Having health personnel that are protected from acquiring the disease during a health emergency is crucial to data collection and management. As such safety of healthcare workers during emergency health situations is crucial.

4.3.7 Case 7: Overlapping Outbreak of COVID-19 and Monkeypox in 2022: Warning for Immediate Preparedness in Iran (Karbalaei & Keikha, 2022)

Background: The human monkeypox viruses circulating in Iran belonged to the B.1 lineage according to the Next strain database. It has been estimated that B.1 originated in Europe and spread around the world (Luna et al. 2022). Although it is assumed that the human monkeypox was introduced to Iran from the southwest neighbouring countries, asymptomatic carriers could play an important role in the human-human transmission chain as natural carriers and reservoirs (De Baetselier et al. 2022).

4.3.7.1 Preparedness and Response Strategies

With no access formulations of tecovirimat and brincidofovir in low-income countries such as Iran, precautions should be taken by the national healthcare authorities. The following were the suggested strategies:

- *Establish a nationwide surveillance* before the disease reaches a critical outbreak situation.
- *Precautionary measures by travellers:* Residents and travelers to endemic areas where there are cases of this disease should observe contact precautions, particularly for MSM which may represent a potential risk of sexual transmission of human monkeypox.
- *Establish counter measures:* Public health facilities should also provide countermeasures such as tracking of contacts and screening to identify and quarantine asymptomatic carriers

The paper suggests immediate preparedness and response strategies to manage the COVID-19 pandemic and monkeypox disease.

4.3.8 Case 8: Lessons Learned from the Ebola Virus Disease and COVID-19 Preparedness to Respond to the Human Monkeypox Virus Outbreak in Low- and Middle-Income Countries (Tusabe et al. 2022)

Background: There are currently no approved treatments for monkeypox virus infections. Given that monkeypox and smallpox viruses are genetically identical, antiviral drugs and vaccines developed to protect against smallpox can be used to prevent and treat infections caused by the monkeypox virus. Antivirals such as tecovirimat (TPOXX) are recommended for immunocompromised patients and others who are more susceptible to severe illness (CDC 2022).

4.3.8.1 Preparedness and Response Strategies

Tusabe et al., suggest the following six core interventions that were used during the Ebola virus disease (EVD) and COVID-19 outbreak to be implemented in response to the Monkeypox outbreak.

- *Establishing an efficient system:* Mathematical disease modeling and diagnostic systems (local/regional) can be used to aid in early detection, including zoonotic spillovers. During the 2014 Ebola outbreak in West Africa, post-outbreak zoonotic niche models demonstrated that the risk of Ebola transmission in West Africa was comparable to that of central Africa, where all previous Ebola outbreaks had occurred (Pigott et al. 2014).
- *Surveillance and contact tracing:* During the Ebola outbreak in the Democratic Republic of the Congo and west Africa, surveillance and

contact tracing were enforced with the participation of national, private, and partner stakeholders. This determined the contact rate between infected and susceptible individuals within distinct populations.

- *Expanding laboratory capacity*: Some LMIC have expanded their laboratory capacities to manage testing for COVID-19 on a large scale using nucleic acid amplification technology (NAAT). Other significant players, such as private and academic laboratories, have supported government initiatives. Likewise, a case from Uganda facilitated the transition to COVID-19 testing by leveraging existing laboratory capabilities developed during Ebola preparedness. The same laboratory capacity could be utilized for diagnosing Monkeypox (Naluyima et al. 2019). For example Nigeria could leverage investments in surge capacity during the immediate wake of the COVID-19 pandemic to expand its functional public health laboratory network from 12 laboratories to 77 fully equipped and functional laboratories capable of testing at varying capacities. SARS Cov2 variant B.1.1.529, 20 K (Omicron) was identified for the first time in South Africa, demonstrating the capabilities of gene sequencing (Oladipo et al. 2020).
- *Prevention, Control and Case Management*: Employing "One Health" approach where Infection, prevention, and control (IPC) guidelines must be developed in collaboration with the ministries of health, animal and water, and the environment. In response to the Ebola virus, countries initiated the IPC cascade, standardizing training curricula and enforcing modifications for resilient health security at entry points, hospitals, and community sites which cabbed the further spread of infections (Aceng et al. 2020).
- *Sustainable infection prevention methods:* Use of various sustainable non-pharmaceutical measures, such as improving hand hygiene practices, and social distancing, which have shown positive results in reducing the rate of infection (Berendes et al. 2022; Zheng et al. 2020).
- *Improved manufacturing in LMIC*: There is need to improve drug manufacturing LMIC. COVID-19 pandemic showed how LMIC had little or no ability to make medical tools, diagnostics, or reagents in-country, and their dependence on import for such crucial products. As borders closed as part of countermeasures to fight the pandemic, most LMIC struggled to get equipment and diagnostic tools it needed to fight COVID-19 (Lee et al. 2020). With 12 vaccine production facilities located in six countries, Algeria, Senegal, Egypt, Morocco, Rwanda, and South Africa, it is anticipated that various COVID-19 vaccines and other antivirals targeting endemic diseases in the region could be produced. Through implementation of partnerships such as the Partnership for African Vaccine Manufacturer (PAVM) framework for Action 2022, technology transfer on mRNA vaccines has been established.

Misinformation and infodemics, misinformation and infodemics within communities could impede intervention efforts in the absence of effective risk communication (Matta 2020). Most LMIC struggled with misinformation which could have contributed to the decision making and health seeking behaviour of communities in the region. Training and making use of trusted leaders such as community and faith leaders as well as other influential people within the society in LMIC could assist in disseminating accurate and up to date health information (Matta 2020).

- *Language of communication*: While some countries have one main language, this is not often the case in some countries as such disseminating health information in the language that is clearly understood by the target community is essential to avoid miscommunication.

Multisectoral approach: Collaboration with various sectors at local, regional, and global level is crucial in combating and implementing counter measures against outbreaks and pandemics. Some countries have mobilized financial resources from different sectors; the National Treasury, multilateral organizations such as the World Bank Group, and bilateral organizations to support the COVID-19 emergency response. Nigeria successfully established a revolving outbreak investigation Fund on the timeliness of outbreak responses in 2019; the fund has supported responses to 14 outbreaks, including Monkeypox.

- *Establishing Regional nodes*: Regional nodes, including emergency operations centers, should be established for stockpiling and prepositioning supplies using a multisectoral approach to ensure timely access to supplies during crises.

Despite the reported successes, there were also reported challenges

- Militarization of public health response in countries such as Kenya, Uganda, and the Democratic Republic of the Congo, during the response to epidemics, this significantly affected community trust.
- Policymaking procedures, misinformation, political will, and governance can be obstacles to transnational cooperation.34 During the response to outbreaks of monkeypox, there is a need to bolster the role of global public health institutions in each nation.
- Financial constraints which limit countries' preparedness and response efforts
- Nature of emerging and or re-emerging disease and transmissibility are crucial in planning for preparedness and response
- Stigma and discrimination

  The paper summarizes some of the achievements and lessons learned from the Ebola disease outbreak and COVID-19 pandemic. There is need for efficient and sustainable systems. Regional and multisectoral collaborations crucial in achieving better preparedness and response to health emergencies. The paper shows that apart from national response, regional collaborations are crucial in combating health emergencies of outbreak prone diseases. Can SDG3d be achieved in this setting? Highly likely but sustainability remains a challenge for many countries in the region.

4.3.9 Case 9: Ebola Virus Disease Preparedness Assessment and Risk Mapping in Uganda, August–September 2018 (Nanziri et al. 2020)

Background: Uganda conducted preparedness and risk-mapping activities to strengthen capacity to prevent EVD importation and spread from cross-border transmission. We adapted the World Health Organization (WHO) EVD Consolidated Preparedness Checklist to assess preparedness in 11 International Health Regulations domains at the district level, health facilities, and points of entry; the US Centers for Disease Control and Prevention (CDC) Border Health Capacity Discussion Guide to describe public health capacity; and the CDC Population Connectivity Across Borders tool kit to characterize movement and connectivity patterns.

4.3.9.1 Epidemic Preparedness and Response Research and Development

- *Diagnosis and Treatment*: Ensure timely availability of diagnostic equipment, treatment, and vaccines required to manage health threats. However, it has been observed that during health threats that these technologies are often limited and not sufficiently available to meet the demand, or they are available but late into the outbreak. Some technologies are ill-adapted and cannot be used in other health systems, hence limiting access.
- *Understanding health technologies*: It is important to understand why and how these health technologies are developed, how they

are deployed, and what makes them fail or succeed.

- *Importance of science communication*: Both the West African EVD epidemic and the COVID-19 pandemic have shown how crucial accurate education and science communication are and how the success of public health interventions depend on it.
- *Misinformation and public health response*: misinformation could undermine epidemic response exposure of the wider population to unfiltered scientific debate—which feeds on diversity of opinions—to public scrutiny, could lead to either naive or malevolent misinterpretations.
- *Failure to filter publications that drive negative opinions:* Scientific journals have also been overwhelmed with submissions, also revealing the limits of the peer review system and its overall capacity to critically absorb a large influx of papers, with sometimes negative consequences on decision-making.
- *Market failure*: Two main factors that have been attributed to market failure are lack of commercial attractiveness of infectious diseases (often requiring short-term therapies, as opposed to more lucrative chronic treatments). As such current health innovation ecosystem which largely rely on the private sector for pharmaceutical product research and development (R&D) and supply may not find infectious diseases attractive for investment.
- *Unequal vaccine access*: Despite the successful development of multiple effective vaccines at unprecedented speed, access to these vaccine remains profoundly unequal, which is a likely engine of persistent transmission and mutations in viral genome.
- *Lack of global coordination:* Globally, coordinated end-to-end R&D ecosystem and a public health-oriented governance mechanism when it comes to the availability and use of interventions is hindering our capacity to effectively prevent and curb epidemics—now and in the future—if radical changes are not set in place.
- *Need for coordination*: There is need for collective intelligence and coordination, not fragmentation and competition between countries and or regions globally.
- *Governments should be able to choose interventions that are applicable, feasible and affective with the existing healthcare system*: Public health authorities would also be expected to (1) define which type of pharmaceutical interventions they need—i.e. determining the target product profile; (2) coordinate a portfolio of R&D projects that covers priority health needs in order to identify and facilitate the development of the most suited products; (3) oversee the research methodology and study designs to ensure they address the most important public health questions in a timely way; and (4) secure the availability, accessibility, and optimal deployment of the health interventions for epidemic preparedness and response.
- *Instituting mechanisms that ensure equitable access of R&D products*: Finally, there is currently no collective mechanism to ensure end-to-end financing of epidemic preparedness and response R&D, including equitable global access. Given that the availability of financing is a critical driver of many activities along the value chain, there is an important role for research funders in directing the right kind of studies and coordination and put conditionalities as needed to the financing to ensure that research efforts are pursued in ways that maximize chances to result in equitable access, including research data sharing.

4.3.9.2 Strengthening the Clinical Research Architecture for Outbreak Diseases Preparedness and Response

- Typically disease outbreaks last from a short period of time making it difficult to test new tools that could be used to combat the disease. Thus research architecture for outbreak prone diseases should consider this in designing preparedness and response strategies.
- Collaborative networks and data sharing is crucial in understanding clinical outcomes as well as other factors pertaining to a particular

disease outbreak. The data could provide insights for future preparedness and response.

Communication, coordination, collaboration, financing, and strong resilient systems are some of the factors that can assist in achieving better preparedness and response to health emergencies. Mapping risks is a crucial step in building a strong and sustainable preparedness and response system.

4.3.9.3 Ensuring Availability and Access

Example of Ebola Virus Disease Vaccines

Of the four new Ebola products that received FDA or EMA marketing authorization, only the rVSV-ZEBOV vaccine (Ervebo) produced by Merck is registered in DRC and in a few other Ebola-prone African countries. It is also WHO prequalified. In January 2021, a GAVI-funded global emergency stockpile of 500,000 doses was created under the auspices of the International Coordinating Group on Vaccine Provision, accessible to all countries.31 While lower income countries can obtain vaccines for free, it is not known how much other countries will be charged, nor the price Merck has charged GAVI. It is also not clear how the supply will be sustained in the future.

Johnson & Johnson's (JNJ) prime-boost vaccine (Zabdeno [Ad26.ZEBOV] and Mvabea [MVA-BN-Filo]) was WHO pre-qualified, is being used in further studies but is not yet registered in African countries. The two treatments, Regeneron's Inmazeb (the association of three monoclonal antibodies: atoltivimab, maftivimab, and odesivimab-ebgn) and Ridgeback Biotherapeutics' Ebanga (the monoclonal antibody ansuvimab), are also not registered in any African country. It is thought that the US government has established stockpiles of at least one, if not both, Ebola treatments, but no details of size and price are publicly available.

Critically, none of these registered products seems readily available for use in disease-endemic countries, including the DRC, which has been experiencing a series of EVD outbreaks. During the 2018–2020 Kivu EVD outbreak—the second largest recorded outbreak—control efforts made use of remaining clinical supplies of the still-investigational products with little transparency on how to access these stocks, and with often restricted conditions for use because still under restrictive "study conditions."

- *The ring vaccination* approach is triggered by an ongoing outbreak, as opposed to vaccinating a population to prevent outbreaks altogether.
- *Lack of efficacy data*, wider use a as preventive vaccine remains questionable, primarily for the lack of clinical efficacy data.
- *Effective Trial designs that benefit endemic populations in the long term*: Here, early trial design choices and unattended evidence gaps now stand in the way of a regional and public health-focused disease control strategy, including targeted preventive vaccination.

Governance, inadequate distribution of ownership, and control: The lack of transparency on conditions attached to the agreements between public and private partners involved in the development of various interventions create gaps in when it comes to decision making and accountability. This affects access not only to the data but also control of manufacturing decisions, local registration, availability and pricing of various vaccines and treatments which most often remains with the commercial partners and not governments.

4.4 Way Forward

Despite the gradual spread of COVID-19 at the onset of the pandemic, many countries could not have anticipated the intensity and impact of the disease on the human population, economy, and infrastructure. Ongoing health campaigns and programs were greatly impacted by the control measures for COVID-19 which limited not only human contact but also led to closures that had significant impact (Use examples of malaria, HIV campaigns, etc.). As such, future preparedness and response plans need to consider management of other diseases both communicable and non-communicable diseases and how they can be sustained in pandemic situations. Sustainable response thus requires a holistic approach that takes into consideration all manner of disease, population, geographic setting, social, economic, and environmental impacts. The COVID-19 pandemic has revealed gaps in health systems, governance, regional and global collaboration, access, and various inequalities that exist. A sustainable response would require therefore addressing these gaps with SDG3d as a guideline.

4.5 Conclusion

Preparedness and response for emerging pathogens of varying characteristics and disease patterns remain crucial in dealing with outbreaks and pandemics. There is need to prepare for social and economic impacts of the disease on society including health personnel, survivors, and caregivers. Among others surveillance, including contact tracing, remains an important factor in identifying cases and contacts although these methods must be improved in most LMIC. Improving health security on the African continent, Latin America as well as LMIC in most parts of the world is needed to combat various public health threats. Upgrading existing infrastructure, improving surveillance and contact tracing, enhancing data sharing and networking including multi sectoral approach could assist in combating future outbreaks. Addressing failures in risk communication and improving communication could assist in reducing misinformation and disinformation. Poverty and syndemics were listed as some of the challenges that have affected preparedness while lack of data, poor governance ad inadequate distribution of equipment, including vaccines and therapeutics and challenges of access and registration of such in endemic countries needs careful consideration. SDG3d is a crucial component in the fight against health emergencies in LMIC and globally. Setting up guidelines according to SDG3d as well as implementation of such, could result in better preparedness and response to health emergency globally.

References

Aceng JR, Ario AR, Muruta AN, Makumbi I, Nanyunja M, Komakech I et al (2020) Uganda's experience in Ebola virus disease outbreak preparedness, 2018–2019. Glob Health 16(1):24. https://doi.org/10.1186/s12992-020-00548-5

Balachandar V, Mahalaxmi I, Subramaniam M, Kaavya J, Senthil Kumar N, Laldinmawii G et al (2020) Follow-up studies in COVID-19 recovered patients – is it mandatory? Sci Total Environ 729:139021. https://doi.org/10.1016/J.SCITOTENV.2020.139021

Berendes D, Martinsen A, Lozier M, Rajasingham A, Medley A, Osborne T et al (2022) Improving water, sanitation, and hygiene (WASH), with a focus on hand hygiene, globally for community mitigation of COVID-19. PLOS Water 1(6):e0000027. https://doi.org/10.1371/journal.pwat.0000027

Brierley L, Vonhof MJ, Olival KJ, Daszak P, Jones KE (2016) Quantifying global drivers of zoonotic bat viruses: a process-based perspective. Am Nat 187(2):E53–E64. https://doi.org/10.1086/684391

CDC. (2022). Treatment information for healthcare professionals. U.S. Department of Health & Human Services: https://www.cdc.gov/poxvirus/monkeypox/clinicians/treatment.html. Accessed 7 Feb 2023

De Baetselier I, Van Dijck C, Kenyon C, Coppens J, Michiels J, de Block T et al (2022) Retrospective detection of asymptomatic monkeypox virus infections among male sexual health clinic attendees in Belgium. Nat Med 28(11):2288–2292. https://doi.org/10.1038/S41591-022-02004-W

Diallo B, Sissoko D, Loman NJ, Bah HA, Bah H, Worrell MC et al (2016) Resurgence of Ebola virus disease in Guinea linked to a survivor with virus persistence in seminal fluid for more than 500 days. Clin Infect Dis 63(10):1353–1356. https://doi.org/10.1093/cid/ciw601

Diallo MSK, Rabilloud M, Ayouba A, Touré A, Thaurignac G, Keita AK et al (2019) Prevalence of infection among asymptomatic and paucisymptomatic contact persons exposed to Ebola virus in Guinea: a retrospective, cross-sectional observational study. Lancet Infect Dis 19(3):308–316. https://doi.org/10.1016/S1473-3099(18)30649-2

El-Shabasy RM, Nayel MA, Taher MM, Abdelmonem R, Shoueir KR, Kenawy ER (2022) Three waves changes, new variant strains, and vaccination effect against COVID-19 pandemic. Int J Biol Macromol 204:161–168. https://doi.org/10.1016/j.ijbiomac.2022.01.118

Endo A, Murayama H, Abbott S, Ratnayake R, Pearson CAB, Edmunds WJ et al (2022) Heavy-tailed sexual contact networks and monkeypox epidemiology in the global outbreak, 2022. Science 378(6615):90–94. https://doi.org/10.1126/science.add4507

European Centre for Disease Prevention and Control. (2022). Mpox (formerly named monkeypox) situation update, as of 17 January 2023. https://www.ecdc.europa.eu/en/news-events/monkeypox-situation-update

Fernandes Q, Inchakalody VP, Merhi M, Mestiri S, Taib N, Moustafa Abo El-Ella D et al (2022) Emerging COVID-19 variants and their impact on SARS-CoV-2 diagnosis, therapeutics and vaccines. Ann Med 54(1):524–540. https://doi.org/10.1080/07853890.2022.2031274

Ferraro F, Caraglia A, Rapiti A, Cereda D, Vairo F, Mattei G et al (2022) Letter to the editor: multiple introductions of MPX in Italy from different geographic areas. Eur Secur 27(23):2200456. https://doi.org/10.2807/1560-7917.ES.2022.27.23.2200456

Galan M, Razzauti M, Bard E, Bernard M, Brouat C, Charbonnel N et al (2016) 16S rRNA amplicon sequencing for epidemiological surveys of bacteria in wildlife. MSystems 1(4):e00032–e00016. https://doi.org/10.1128/mSystems.00032-16

Happi C, Adetifa I, Mbala P, Njouom R, Nakoune E, Happi A et al (2022) Urgent need for a non-discriminatory and non-stigmatizing nomenclature for monkeypox virus. PLoS Biol 20(8):e3001769. https://doi.org/10.1371/journal.pbio.3001769

Haque A, Pant AB (2022) Mitigating Covid-19 in the face of emerging virus variants, breakthrough infections and vaccine hesitancy. J Autoimmun 127:102792. https://doi.org/10.1016/j.jaut.2021.102792

Harris RC, Chen Y, Côte P, Ardillon A, Nievera MC, Ong-Lim A et al (2021) Impact of COVID-19 on routine immunisation in South-East Asia and Western Pacific: disruptions and solutions. Lancet Reg Health West Pacific 10:100140. https://doi.org/10.1016/j.lanwpc.2021.100140

Hussain S, Hussain A, Ho J, Sparagano OAE, Zia U-R (2020) Economic and social impacts of COVID-19 on animal welfare and dairy husbandry in Central Punjab, Pakistan. Front Vet Sci 7:589971. https://doi.org/10.3389/fvets.2020.589971

Iglesias Martínez E, Roces García J, Jiménez Arberas E, Llosa JA (2022) Difference between impacts of COVID-19 on women and men's psychological, social, vulnerable work situations, and economic well-being. Int J Environ Res Public Health 19(14):8849. https://doi.org/10.3390/ijerph19148849

Izudi J, Kiragga AN, Kalyesubula P, Okoboi S, Castelnuovo B (2022) Effect of the COVID-19 pandemic restrictions on outcomes of HIV care among adults in Uganda. Medicine 101(36):e30282. https://doi.org/10.1097/MD.0000000000030282

Jones KE, Patel NG, Levy MA, Storeygard A, Balk D, Gittleman JL, Daszak P (2008) Global trends in emerging infectious diseases. Nature 451(7181):990–993. https://doi.org/10.1038/nature06536

Karbalaei M, Keikha M (2022) Overlapping outbreak of COVID-19 and monkeypox in 2022: warning for immediate preparedness in Iran. Int J Surg 105:106892. https://doi.org/10.1016/j.ijsu.2022.106892. Epub 2022 Sep 6. PMCID: PMC9444578.

Khankeh H, Farrokhi M, Roudini J, Pourvakhshoori N, Ahmadi S, Abbasabadi-Arab M et al (2021) Challenges to manage pandemic of coronavirus disease (COVID-19) in Iran with a special situation: a qualitative multi-method study. BMC Public Health 21(1):1–9. https://doi.org/10.1186/s12889-021-11973-5

Klinton JS, Heitkamp P, Rashid A, Faleye BO, Win Htat H, Hussain H et al (2021) One year of COVID-19 and its impact on private provider engagement for TB: a rapid assessment of intermediary NGOs in seven high TB burden countries. J Clin Tuberc Other Mycobact Dis 25:100277. https://doi.org/10.1016/J.JCTUBE.2021.100277

Lee VJ, Aguilera X, Heymann D, Wilder-Smith A, Lee VJ, Heymann DL et al (2020) Preparedness for emerging epidemic threats: a lancet infectious diseases commission. Lancet Infect Dis 20(1):17–19. https://doi.org/10.1016/S1473-3099(19)30674-7

Li Y-T, Linster M, Mendenhall IH, Su YCF, Smith GJD (2019) Avian influenza viruses in humans: lessons from past outbreaks. Br Med Bull 132(1):81–95. https://doi.org/10.1093/bmb/ldz036

Luna N, Ramírez AL, Muñoz M, Ballesteros N, Patiño LH, Castañeda SA et al (2022) Phylogenomic analysis of the monkeypox virus (MPXV) 2022 outbreak: emergence of a novel viral lineage? Travel Med Infect Dis 49:102402. https://doi.org/10.1016/J.TMAID.2022.102402

Madzima B, Makoni T, Mugurungi O, Mudariki G, Mpofu A, Dube F et al (2022) The impact of the COVID-19 pandemic on people living with HIV in Zimbabwe. Afr J AIDS Res 21(2):194–200. https://doi.org/10.2989/16085906.2022.2103004

Matta G (2020) Science communication as a preventative tool in the COVID19 pandemic. Humanit Soc Sci Commun 7(1):159. https://doi.org/10.1057/s41599-020-00645-1

McClarty L, Lazarus L, Pavlova D, Reza-Paul S, Balakireva O, Kimani J et al (2022) Socioeconomic burdens of the COVID-19 pandemic on LMIC populations with increased HIV vulnerabilities. Curr HIV/AIDS Rep 19:76–85. https://doi.org/10.1007/s11904-021-00591-w

Monkeypox. (n.d.). https://www.who.int/news-room/fact-sheets/detail/monkeypox. Accessed 25 Jan 2023

Murray KA, Preston N, Allen T, Zambrana-Torrelio C, Hosseini PR, Daszak P (2015) Global biogeography of human infectious diseases. Proc Natl Acad Sci U S A 112(41):12746–12751. https://doi.org/10.1073/pnas.1507442112

Naluyima P, Kayondo W, Ritchie C, Wandege J, Kagabane S, Tumubeere L et al (2019) The joint mobile emerging disease clinical capability (JMEDICC) laboratory approach: capabilities for high-consequence pathogen clinical research. PLoS Negl Trop Dis 13(12):e0007787. https://doi.org/10.1371/JOURNAL.PNTD.0007787

Nanziri C, Ario AR, Ntono V, Monje F, Aliddeki DM, Bainomugisha K et al (2020) Ebola virus disease preparedness assessment and risk mapping in Uganda, august-September 2018. Health Secur 18(2):105–113. https://doi.org/10.1089/hs.2019.0118

Nkengasong J, Djoudalbaye B, Maiyegun O (2017) A new public health order for Africa's health security. Lancet Glob Health 5:e1064–e1065. https://doi.org/10.1016/S2214-109X(17)30363-7

OECD. (2022) First lessons from government evaluations of COVID-19 responses: A synthesis. https://read.oecd-ilibrary.org/view/?ref=1125_1125436-7j5hea8nk4&title=First-lessons-from-government-evaluations-of-COVID-19-responses

Oladipo EK, Ajayi AF, Odeyemi AN, Akindiya OE, Adebayo ET, Oguntomi AS et al (2020) Laboratory diagnosis of COVID-19 in Africa: availability, challenges and implications. Drug Discov Ther 14(4):153–160. https://doi.org/10.5582/DDT.2020.03067

Peel AJ, Wells K, Giles J, Boyd V, Burroughs A, Edson D et al (2019) Synchronous shedding of multiple bat paramyxoviruses coincides with peak periods of Hendra virus spillover. Emerg Microbes Infect 8(1):1314–1323. https://doi.org/10.1080/22221751.2019.1661217

Perez Duque M, Ribeiro S, Martins JV, Casaca P, Leite PP, Tavares M et al (2022) Ongoing monkeypox virus outbreak, Portugal, 29 April to 23 May 2022. Euro Surveill 27(22):2200424. https://doi.org/10.2807/1560-7917.ES.2022.27.22.2200424

Petersen E, Zumla A, Hui DS, Blumberg L, Valdoleiros SR, Amao L, Ntoumi F, Asogun D, Simonsen L, Haider N, Traore T, Kapata N, Dar O, Nachega J, Abbara A, Al Balushi A, Kock R, Maeurer M, Lee SS, Lucey DR, Ippolito G, Koopmans M (2022) Vaccination for monkeypox prevention in persons with high-risk sexual behaviours to control ongoing outbreak of monkeypox virus clade 3. Int J Infect Dis 122:569–571. https://doi.org/10.1016/j.ijid.2022.06.047. Epub 2022 Jul 2. PMCID: PMC9534076

Pigott DM, Golding N, Mylne A, Huang Z, Henry AJ, Weiss DJ et al (2014) Mapping the zoonotic niche of Ebola virus disease in Africa. elife 3:e04395. https://doi.org/10.7554/ELIFE.04395

Pillay Y, Pienaar S, Barron P, Zondi T (2021) Impact of COVID-19 on routine primary healthcare services in South Africa. S Afr Med J 111(8):714–719. https://doi.org/10.7196/SAMJ.2021.V111I8.15786

Queeneth Ojoma A, David O, Chukwuma AC, Daniel Ugochukwu N, Godwin M, Ezike C (2021) Transmission dynamics of Ebola virus disease with vaccine, condom use, quarantine, isolation and treatment drug. J Infect Dis 15(1):10–23. https://doi.org/10.21010/ajid

Rodríguez BS, Herrador BRG, Franco AD, Fariñas MPS-S, Valero J d A, Llorente AHA et al (2022) Epidemiologic features and control measures during monkeypox outbreak, Spain, June 2022. Emerg Infect Dis 28(9):1847. https://doi.org/10.3201/EID2809.221051

Rodriguez-Morales AJ, Lopardo G, Verbanaz S, Orduna T, Lloveras S, Azeñas-Burgoa JM et al (2022) Latin America: situation and preparedness facing the multi-country human monkeypox outbreak. Lancet Reg Health Am 13:100318. https://doi.org/10.1016/j.lana.2022.100318

Shirani K, Sheikhbahaei E, Torkpour Z, Ghadiri Nejad M, Kamyab Moghadas B, Ghasemi M et al (2020) A narrative review of COVID-19: the new pandemic disease. Iran J Med Sci 45(4):233–249. https://doi.org/10.30476/ijms.2020.85869.1549

Subissi L, Keita M, Mesfin S, Rezza G, Diallo B, Van Gucht S et al (n.d.) Ebola virus transmission caused by persistently infected survivors of the 2014-2016 outbreak in West Africa. J Infect Dis 2018(5):287. https://doi.org/10.1093/infdis/jiy280

Tamang ST, Dorji T (2022) Bhutan's preparedness for monkeypox outbreak. Lancet Reg Health Southeast Asia 7:100092. https://doi.org/10.1016/j.lansea.2022.100092

Turk MA, Mitra M (2021) COVID-19 and people with disability: social and economic impacts. Disabil Health J 14(4):101184. https://doi.org/10.1016/j.dhjo.2021.101184

Tusabe F, Tahir IM, Akpa CI, Mtaki V, Baryamujura J, Kamau B et al (2022) Lessons learned from the Ebola virus disease and COVID-19 preparedness to respond to the human Monkeypox virus outbreak in low- and middle-income countries. Infect Drug Resist 15:6279–6286. https://doi.org/10.2147/IDR.S384348

UK Government. (2022). Monkeypox cases confirmed in England – latest updates. https://www.gov.uk/government/news/monkeypox-cases-confirmed-in-england-latest-updates

Van Kerkhove MD, Bento AI, Mills HL, Ferguson NM, Donnelly CA (2015) A review of epidemiological parameters from Ebola outbreaks to inform early public health decision-making. Sci Data 2:150019. https://doi.org/10.1038/sdata.2015.19

WHO. (2022a) Multi-country monkeypox outbreak: situation update. WHO https://www.who.int/emergencies/disease-outbreak-news/item/2022-DON396. Accessed 23 Jan 2023

WHO. (2022b) Vaccines and immunization for monkeypox. Who, (June), 1–28. https://www.who.int/publications/i/item/who-mpx-immunization-2022.1

Xie Y, Xu E, Al-Aly Z (2022) Risks of mental health outcomes in people with covid-19: cohort study. BMJ:e068993. https://doi.org/10.1136/bmj-2021-068993

Xu G, Liu F, Ye M, Zhao J, Li Q, Feng C et al (2020) No evidence of re-infection or person-to-person transmission in cured COVID-19 patients in Guangzhou, a retrospective observational study. Front Med 7:593133. https://doi.org/10.3389/fmed.2020.593133

Yinka-Ogunleye A, Aruna O, Dalhat M, Ogoina D, McCollum A, Disu Y et al (2019) Outbreak of human monkeypox in Nigeria in 2017–18: a clinical and epidemiological report. Lancet Infect Dis 19(8):872–879. https://doi.org/10.1016/S1473-3099(19)30294-4

Zheng Q, Jones FK, Leavitt SV, Ung L, Labrique AB, Peters DH et al (2020) HIT-COVID, a global database tracking public health interventions to COVID-19. Sci Data 7(1). https://doi.org/10.1038/S41597-020-00610-2

Zumla A, Valdoleiros SR, Haider N, Asogun D, Ntoumi F, Petersen E, Kock R (2022) Monkeypox outbreaks outside endemic regions: scientific and social priorities. Lancet Infect Dis 22(7):929–931. https://doi.org/10.1016/S1473-3099(22)00354-1

Achieving SDG3.3 and 3d in the Era of Misinformation

5

Abstract

Health information is a very important as of our world today. There are so many reasons why people use and or need health information around the world. It could be to gather information on health issues such as disease, vaccines, and therapies but also to learn more about what is going on around and stay up to date regarding their health and wellbeing. Apart from obtaining the right information, frequent use and dependence on social media has seen the growth of misinformation which has negatively affected health outcomes in various ways. Making informed decisions is one of the crucial factors in good health and wellbeing, clinical research as well as public health; having the right information is therefore crucial. With infectious disease, this can make huge difference in preparedness, response, and disease outcomes. Low-and-middle income countries are lagging in technological advancements and are often equipped with little information and or misinformed in one way or the other. How can the public benefit from information on health and research outcomes? Apart from research papers and publication, how can the scientific community make use of social media and other media platforms to inform the Governments and the public to combat misinformation? Is there another way to reach populations that do not have access to clinical and scientific research?

Keywords

Health information · Misinformation · Malinformation · Disinformation · SDG3 · Social media · Ehealth · Health literacy · Ehealth literacy · Media health literacy

5.1 Misinformation

Health information is a very important as of our world today. There are so many reasons why people use and or need health information around the world. It could be to gather information on health issues such as disease, vaccines, and therapies but also to learn more about what is going on around them regarding their health and wellbeing. With the ongoing pandemic and outbreaks a lot of health related information has been circulating including come which were incorrect, utterly false and or misleading. This phenomenon known as misinformation has implications on health outcomes.

According to Chou et al., "health misinformation" can be defined as any health-related claim of fact that is false based on current scientific consensus (Sylvia Chou et al. 2020). There are two types of misinformation, false information but not created with intention to cause harm (mis-

F. A. Mphande, *Sustainable Health in Low and Middle Income Countries*, Sustainable Development Goals Series, https://doi.org/10.1007/978-981-99-4254-1_5

information), and false information based on reality but created to harm a particular person, social group, institution or country (disinformation/malinformation) (Suarez-Lledo and Alvarez-Galvez 2021) (Fig. 5.1).

The unbridled use of social media has encouraged the spread of misinformation allowing it to reach many around the world. As information from one individual to the next it is often distorted and may partially and or completely lose its original meaning. Misinformation through social media has become a concerning public health issue calling for the need to regulate the quality and availability of public health information online (17, 18). Despite the knowledge of existing misinformation there is little evidence on the effect of such practice on communities worldwide hence the need for further research.

5.1.1 Education, Gender Equity and Misinformation

Education is crucial in every society around the world. Literacy rate has increased in the past decades, in 2020, the literacy rate for adults (15 years and above) globally was 87% compared to 67% and 81% in the year 1976 and 2000 respectively. (https://data.worldbank.org/indicator/SE.ADT.LITR.ZS?view=chart). Despite this increase in literacy rate, challenges remain in LMIC and the global South. Even though one is able to read and write, the quality of education is crucial in understanding the various policies, innovations and technologies. The twenty-first century has ushered in a new era of digital advancement which is developing at lightning speed. While the global North is able to keep up with these advancements, the global South remains far behind (Sepúlveda et al. 2022). Sustainable development goal 4 (SDG4) aims to "ensure inclusive and equitable quality education and promote lifelong learning opportunities for all." In the global South as are most LMIC social exclusion, gender inequality, politics, religion, and poverty remain a hindrance for children to attain a good education (Chandra 2022; Sepúlveda et al. 2022). This gap in knowledge has paved way for misinformation in populations with limited or no education with limited ability to distinguish between facts and falsity. Education is an important aspect in attaining SDG3. SDG3.3 and 3d require understanding of various diseases, their management and control, as such there is need for basic education within communities. If

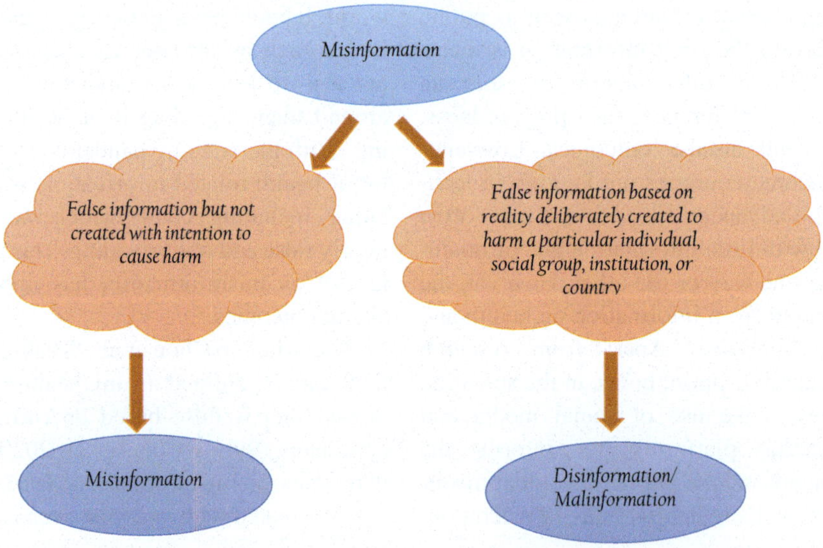

Fig. 5.1 Types of misinformation

communities are able to acquire and comprehend basic health information, and attain basic health literacy, there is a greater chance in improving health outcomes.

Socially, various roles, relationships, personality traits, attitudes, behaviours, values, and power have been attributed to the two sexes, men and women on a differential basis. By comparing the relative roles between the sexes and the outcomes, inequalities have been reported. According to UNESCO, Women constitute approximately two thirds (514 million) of adults without basic literacy skills (https://www.unesco.org/en/gender-equality/education#:~:text=Gender%20equality%20is%20a%20global,boys%20are%20out%20of%20school). Gender plays an important role in various health policies and health related outcomes (Chikovore et al. 2015; Iglesias Martínez et al. 2022; Meites et al. 2016; Mousley et al. 2014). Subservience of women to men in some cultures has affected decision making on womens' reproductive health as well as other health related decision (Vlassoff 2007). The COVID-19 pandemic further affected education for many globally with intermittent school closures which resulted millions of children dropping out of school (Afzal et al. 2022; Nuwematsiko et al. 2022). Education and gender are some of the determinants that affect health literacy and play a role in health related misinformation (Budhathoki et al. 2017; Gibbs et al. 2012).

5.2 Medical and Health-Related Misinformation on Social Media

The internet has and continues to be a place where people find information. With the advent of smartphones, the world is now more connected than ever before. The growing social media trend has also seen growth of information sharing globally. The information shared is often poorly regulated, leading to the spread of both right and wrong information. While the right information is good for the community, wrong information can be misleading and in the case of health information, it is a thin line which could build or break health campaigns globally with concerning outcomes (Caballero-Anthony 2021; Hernandez et al. 2021; Jaiswal and Halkitis 2019; Mack et al. 2014; SteelFisher et al. 2015). For example, growth of misinformation regarding vaccines has seen an increase in vaccine hesitancy which has seriously affected campaigns for vaccine preventable diseases (Abu-Odah et al. 2022; Borges do Nascimento et al. 2022; Hernandez et al. 2021).

Apart from the misinformation, limited health literacy has contributed to further exacerbate the problem (Amoah 2018; Bresolin 1999; Cohen et al. 2015; Gibbs 1996; Shahid et al. 2022; Sørensen et al. 2012; Stormacq et al. 2020). Those with better understanding of health issues can sift through health information online and can identify false information from the truth. Contrary, those with limited health literacy have challenges to do so. In LMIC poor health literacy and misinformation has led to loss of trust on governments and health authorities, poor reception, and participation in health campaigns as well as lack of willingness and adherence to health interventions and regulations (Jaiswal and Halkitis 2019; Tasnim et al. 2020).

Individuals are more likely to spread negative news compared to positive news. A study by on link between MMR and autism has seen ripples effects on vaccine hesitancy even after the paper was retracted (A Timeline of the Wakefield Retraction 2010; "Wakefield's article linking MMR vaccine and autism was fraudulent," Godlee et al. 2011; Dyer 2010; Omer 2020).

Health literacy is one tool that could utilised to reduce knowledge gaps and assist society in identifying misinformation from truth. Health literacy (HL) is conceptualized as skills and competences enabling people to obtain and interpret health information and apply their knowledge to inform health-related decision-making (Sørensen et al. 2012). Health literacy can further be divided into two closely linked but distinct concepts Media Health Literacy (MHL) and eHealth Literacy (eHL) (Levin-Zamir et al. 2011; Norman and Skinner 2006).

Limited health literacy could affect peoples' understanding of MHL and eHL. There are four gaps of knowledge that have been identified which contribute to the spread of misinformation. The first one dominant misinformation trends, followed by opportunity to increase health literacy, self-efficacy, and finally self-treatment. To reduce and or eliminate these gaps, there is need to understand who is seeking this health information, where are they seeking this information and why they are seeking this information. Health information is sought by people from all backgrounds, with varying literacy levels and needs. For example, the reason why health professionals may seek information and how they attain this information can be different from the way the wider population will seek similar type of information. The places where this information is sought, and the quality of the information may also differ depending on who is sharing the information and how much in-depth knowledge the sharer possesses (Fig. 5.2). While social media can be used as a vehicle to spread information across communities around the world, some of the information spread through this medium is partially correct or false leading to misinformation and or disinformation. Similarly, health professionals may use social media and other health media platforms to spread information related to health and disease, but the message can be misunderstood, tweeted and or misreported leading to misinformation (Fig. 5.2).

5.2.1 Media Health Literacy (MHL) and Electronic Health Literacy (eHL)

Nowadays information can is available in different forms (digital and non- digital media) and can be accessed from various sources including TV, radio, internet, mobile phones. All these sources provide all kinds of information including dissemination of health information, disease prevention and control campaigns, information on outbreaks as well as updates on disease distribution and spread within a particular community; poverty, lack of resources, and illiteracy affect access to information(Fig. 5.3).

5.2.2 Determinants of MHL and eHL

Although MHL and eHL are widely being used in the twenty-first century, not everyone is able to utilize these media sources. There are several factors that have been reported as determinants of MHL and eHL. These include environmental factors that could be linked to social, economic, and organizational context. The social environment, networks and how people interact is a good source of information, whether the information is credible or not is another point of discussion. Economically, having resources can allow an individual to access all kinds of media both MHL and eHL. Ability to buy or access electronic equipment and gadgets, access to internet and the skills to use these gadgets determines how much information an individual, community and or society can be exposed to.

Apart from skills, individuals should be able to understand and interpret the information being provided. With MHL, communications should be able to offer health guidance to all with clarity to reach the desired population. Attention should be paid to mass media content generated for commercial entities and health organization so that the content is health promoting and not health compromising. The population should be able to identify implicit and explicit health information from various types of media and recognize its influence on health behaviour. It has been reported that health outcomes are associated with health literacy, and these include health behaviour and health status (DeWalt et al. 2004; Shahid et al. 2022; Stormacq et al. 2020). Individuals should be able to critically analyze content and the intentions thus having the ability to take measured response.

5.2.2.1 Media Health Literacy

Media health literacy is the precursor to eHealth literacy. "Electronic health literacy has been defined as the ability to seek, find, understand and approve health information from electronic

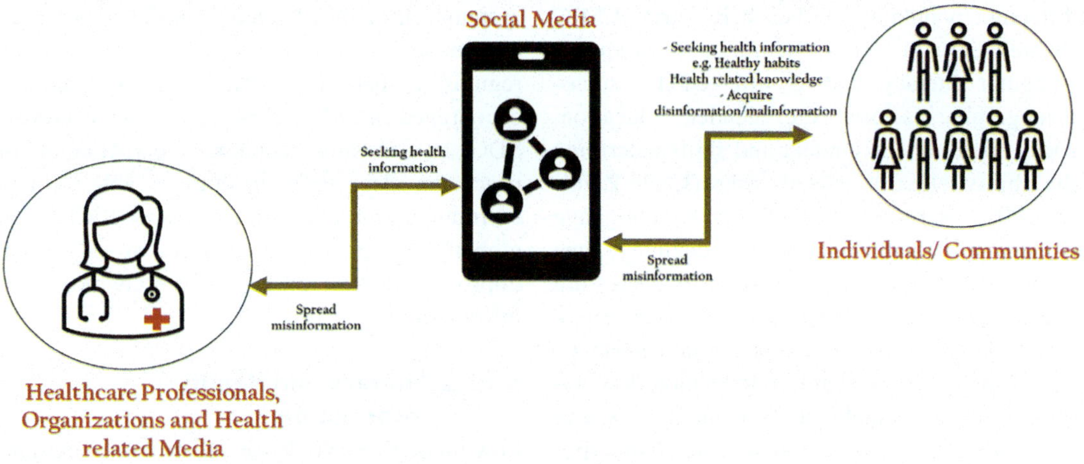

Fig. 5.2 Spread of health information

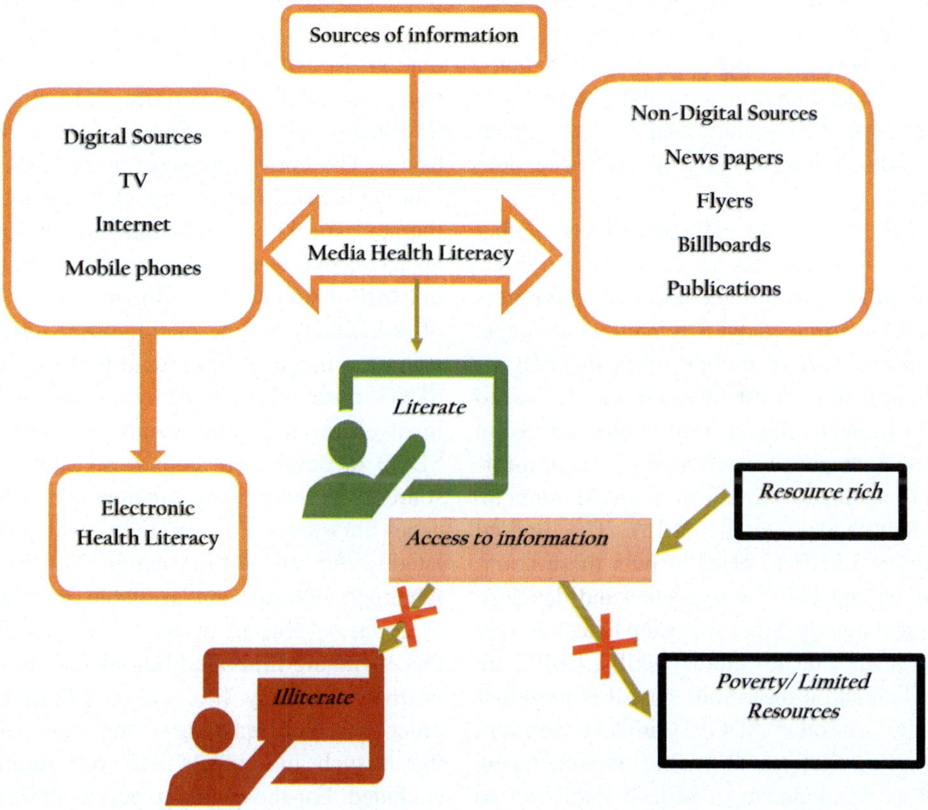

Fig. 5.3 Media health literacy and electronic health literacy

sources and apply knowledge gained to addressing and solving problems (Norman and Skinner 2006). eHealth literacy is can be viewed as combination of six literacy domains: traditional literacy, information literacy, scientific literacy, media literacy, computer literacy and health literacy (Norman and Skinner 2006). Several barriers for eHealth exist and have been identified in different populations elsewhere and LMIC (Schreiweis et al. 2019). eHealth literacy requires computer literacy and is affected by socio-demographic factors. Age, gender, education, social environment, context and skills determine the ability eHealth related tasks (Levin-Zamir et al. 2011; Norman and Skinner 2006). Language barriers were identified as another challenge, one is only able to use electronic services if they are able to read, write and understand the language in which the information is being relayed. Hence there is need for translating information into languages that are available in the context and population where the information is being shared. For example many countries in Africa have more than one language as such language barriers are a huge determinant of MHL and eHL. Language translations are thus required. Sharing information in a more interactive format also helps enhance understanding. Content should be adopted to specific population taking into consideration culture, habits, and context (Feuerstein 1999).

In LMIC MHL and eHL are advancing but gaps still remain. Access to electricity and internet, economic hardships and access to computers and other electronic gadgets remain a challenge. While the number of mobile users in LMIC is high, there is limited use of electronic devises to seek and or access health information (Bresolin 1999). Social media is a common source of information including health information (Marar et al. 2019; Smailhodzic et al. 2016). The limited exposure of LMIC to eHealth puts these countries far behind HICs in exposure and development of technology. Since exposure to technology strongly shapes health literacy skills, LMIC are lagging behind on these skill sets. It is therefore crucial to consider SDG9, building resilient infrastructure that would enable manufacturing as well as development of hi-tech industries to meet the needs of populations in in LMIC. Development in LMIC has been hindered by several factors including health emergencies, natural disasters, corruption, and political instability. Political instability has resulted in disruption of health services, including immunization and routine public health programmes and health systems within the affected countries. Peace is an essential asset for prosperity within a country and regions, as such SDG16 has to be promoted and encouraged in LMIC if SDG3 is to be achieved. SDG 16 not only encourages perpetuation of peace but also strong institutions able to fight corruption and other injustices in LMIC and globally. Resilient infrastructures in LMIC will improve eHL through access to technology and information.

5.2.2.2 Literacy and Health Information

Low literacy rates in LMIC as well as in marginalized communities in HICs has been reported to be a hindrance to success of health programmes (Amoah 2018; Irwan et al. 2016; Kabir and Afzal 2016). With low literacy, it is a challenge for populations to distinguish truth from false information as such availability of unregulated information on the internet is a threat to public health. The current move to accessing information via internet has created challenges in health literacy especially in populations with limited or no access to internet. With most of the population in LMIC unable to access the internet, this group of individuals will be marginalized further widening the inequality gaps and making achieving SDG3 difficult. Popoola et al., proposed that involvement of libraries could assist in achieving SDG3 in developing countries. Libraries are a source of credible information and health sciences libraries could assist in disseminating regulated health related information (Popoola 2019). Although libraries are available in LMIC, not everyone is able to access these facilities. For those who are illiterate libraries may not be part of their everyday life, unless libraries devise other ways of reaching communities such as these, such individuals will be automatically excluded. For those able to access libraries their level of health literacy will determine how much of the available health related information they can understand. Use of various modes of communication including audio and visual communication such as infographics are thus essential in reaching out to a wider audience (Arcia et al.

2019; Fagerlin et al. 2017; Nematollahi et al. 2020; Traboco et al. 2022; Van Hecke et al. 2020).

Mobile libraries could be another option of reaching communities with limited or no access to libraries. With varying levels of health systems in LMIC, countries could adopt a system of improving health literacy within communities through mobile libraries that could reach populations countrywide. Popoola proposes that libraries in developing countries can counteract the effect of the low resource status of their healthcare system on health literacy by proactively facilitating access to quality information, literacy skills and lifelong learning for individuals, community and practitioners (Popoola 2019). The authors suggest that health sciences libraries can be relevant in a LMIC health care system and SDG-3 plans by working together to share successful and failed initiatives implemented towards enhanced health literacy for all (Popoola 2019). These mobile libraries could collaborate with various health professionals to devise ways and means to disseminate information both to those who are literate and illiterate so that no one is left behind. The information could be made available for all ages so that the various age groups are reached and equipped with the required and relevant health information. Through collaboration with health professionals, librarians could use their knowledge and skills to create media that can reach out populations within their countries considering, the demographics, culture, habits, health system, environment, social aspects, and setting.

While other modes of communication such as libraries and infographics could be helpful in disseminating information, the ability to understand the information differs. The difference has been attributed to several factors these include personal, situational, and environmental factors (DeWalt et al. 2004; Norman and Skinner 2006). Personal factors include the age, gender, social economic factors, ethnicity, and culture, these are also known as social demographic factors. Individual skills such as literacy, language barriers, as well as technological experience also affect personal understanding of health information. Another personal aspect to consider that affects understanding of health information is perception. Depending on the knowledge, literacy, skills, culture, age, gender including social economic factors, individuals perceptions are built. These perceptions affect understanding as well as decision making concerning their health and wellbeing. With low literacy rates, limited access, marginalization, culture, poverty/limited resources, and inequalities that exist in LMIC, the perceptions towards health challenges and interventions may not be comparable to other countries. Secondly, situational factors play a role in how people understand health information. Emotionally, people who have experienced a disease, either suffering from the disease of caring for someone suffering from a particular disease, these individuals are more aware of the health threats and may be cautious about their health status compared to those without experience. Emotionally such individuals may react differently to health information and disease threat. Social economic status is another situational aspect that plays a role in understanding of health information. While individuals with higher economic status can access both digital and nondigital media, this group is often well informed compared to those with low social economic status. Thus, the understanding of health information could also differ (Jaiswal and Halkitis 2019). Past experiences, as well as environmental factors such as access to technology, relevance, and appropriateness to target audience including social support are some of the aspects that contribute to understanding and openness to various health interventions and information. Past experiences including exclusion, discrimination, racism, and other health disparities particularly in vulnerable populations and ethnic minorities have led to loss of trust in health systems and health information. As such these groups may not be as open to health information even if they have access to such information (American Psycholgical Association: APA Working Group on Health Disparities in Boys and Men 2018; Boulware et al. 2003; Mukherjee 2020; Selvarajah et al. 2022; Shannon et al. 2022; Thorburn et al. 2012).

A study in Nepal reported that some of the barriers in healthcare engagement in the country include, knowledge and education, culture, gender roles, quality of service, cost of service, as well as health literacy (Budhathoki et al. 2017). The study pointed out health literacy as a determinant of understanding, access and use of health information and services. As such better understanding of the health literacy needs for a population is required to address these needs effectively (Budhathoki et al. 2017). Identification of local needs and developing health literacy interventions that are needed within the population could provide opportunities for systemic improvements within the healthcare system (Budhathoki et al. 2017). This idea could be beneficial in addressing health literacy in LMIC globally. While there is ample data on use of eHealth tools among populations worldwide the data on ethnic minorities is limited. It is therefore crucial when reading data on Media and eHealth literacy as interventions to consider equity, access, cultural appropriateness, overcoming the digital divide, and the various stages of development of health systems globally. While MHL and eHL is gaining ground in many parts of the world, some countries are still far behind. As such SDG4 (Education), SDG9 (resilient infrastructure) and SDG16 (peace) are crucial for development of better health systems globally and ensuring that no one is left behind in attaining SDG3.

5.3 COVID-19 and Misinformation

The coronavirus disease 2019 (COVID-19) pandemic has not only caused significant challenges for health systems all over the globe but also fuelled the surge of numerous rumours, hoaxes, and misinformation, regarding the aetiology, outcomes, prevention, and cure of the disease. The interventions to prevent the spread of COVID-19 included mass lockdowns which affected businesses and people from all walks of life. The lockdowns affected both the social and economic aspects of peoples' lives resulting in loss of income, loss of physical interaction and mental health issues. These changes saw the growth of online shopping and increased interactions through the internet and social media (Vall-Roqué et al. 2021). The use of social media led to exposure to a high volume of information lead to media fatigue, and discontinuation of healthy behaviours that are essential to protect individuals. Furthermore, misinformation and rumours regarding COVID-19 hindered the practice of healthy behaviours (such as handwashing and social distancing) and promoting erroneous practices that increase the spread of the virus and ultimately resulting in poor physical and mental health outcomes (Tasnim et al. 2020). Presentation of information by mainstream media can also contribute to misinformation. The way information is initially presented by mainstream mass media affects how this information may be interpreted by other media platforms. Information is often exaggerated and misrepresented on social media platforms leading to local and global reaction often with negative outcomes including discrimination of minorities, exclusion, as well as abuse (Bora et al. 2018; Dionne and Turkmen 2020). The stress associated with COVID-19 sometimes led to extreme reactions and outcomes e.g. Suicide, as well as drug overdose (Banerjee et al. 2021; Farooq et al. 2021; The Lancet Psychiatry 2021). The COVID-19 crisis has shown that diseases which affect minority groups can activate negative reactions including xenophobia towards those groups (Reny and Barreto 2022). Research has shown that disease outbreaks have been associated with blaming minority and or vulnerable groups for infectious diseases outbreaks which have often led to policies that discriminate such groups (Dionne and Turkmen 2020; Sotgiu and Dobler 2020). These negative responses play a crucial role as to how such groups respond to global health interventions, health information and policies (Abu-Odah et al. 2022; Dionne and Turkmen 2020; Sotgiu and Dobler 2020). The global reaction to the news of COVID-19 omicron variant, led to closure of borders affecting travels for many coming from Sub Saharan Africa and negative reactions to citizens coming from the region globally. The question remains with such attitudes, where there is a distinction of "us" and "them" how can sustainable health be achieved?

5.3.1 Prevalence of Misinformation

A recent review revealed that health information on social media was a concern, with misinformation from social media posts regarding vaccines up to 51%, approximately 60% related to pandemics and around 28% of posts on COVID-19 (Borges do Nascimento et al. 2022). Up to 30% of videos on YouTube on emerging infectious diseases were reported to be inaccurate and misleading. https://www.who.int/europe/news/item/01-09-2022-infodemics-and-misinformation-negatively-affect-people-s-health-behaviours%2D%2Dnew-who-review-finds

It was observed that misinformation had effect on mental, social, political and economic distress. Despite the negative effects, there was a significant improvement on knowledge awareness and positive health-related behaviours in some circles (Borges do Nascimento et al. 2022).

The prevalence of health misinformation was the highest on Twitter and on issues related to smoking products and drugs. However, misinformation on major public health issues, such as vaccines and diseases, was also high (Borges do Nascimento et al. 2022).

Misinformation is often associated with knowledge gaps, as these knowledge gaps widen, people tend to seek for answers to lingering question (Suarez-Lledo and Alvarez-Galvez 2021). With the wide use of social media, this is where information is easily sought and found including inaccurate information (Sylvia Chou et al. 2020). To find ways how to critically reduce misinformation, it is important to understand why misinformation is widely spread on social media platforms. Misinformation tends to follow trending topics on social media as such it is crucial to identify the dominant misinformation trends on the various social media platforms. The interactive aspects of various social media platforms have also contributed to the spread of misinformation. As people chase for more "likes" and "views" factors are highly likely to encourage the progressive spread health misinformation. But how does misinformation affect public health? Despite the limited amount of data on how misinformation has affected public health, several factors have been observed and misinformation is suggested as a likely contribute to these outcomes. Vaccine hesitancy is one example that widely observed during the COVID-19 pandemic as well as in the emergence of vaccine preventable diseases in areas where the vaccine is available, but individuals are less likely to take the vaccine due to their beliefs or other reasons (Dredze et al. 2016). The widespread misinformation and politicization of vaccine distribution further strengthened the myths and conspiracies, further contributing to vaccine hesitancy and lack of compliance to health information. Thus when designing health strategies, consideration should be taken on how misinformation can be reduced and how misinformation may affect and or impact the interventions (Aylward and Tangermann 2011). There is need to understand the roots of vaccine hesitancy and refusal and find better ways to tackle this challenge. Hernandez et al. suggest use of healthcare professionals as an academically and medically valuable resource, to share the right kind of information on social media to combat infodemics (Hernandez et al. 2021). It is crucial that healthcare and academic organization should be on the forefront to distribute safeguard information regarding health and health related topics to ensure that misinformation is tackled effectively. Hernandez et al., suggested the following strategies. Developing and implementing internal methods to detect prevalent health misinformation; requiring and expanding conversations with social media executives to identify safeguards against health-care-related misinformation; delegating socio-culturally diverse frontline staff and HCPs with roles specific to pro-vaccination to counter COVID-19 misinformation by sharing their stories and perspectives; incentivizing and supporting HCPs within organizations to be active on social media; and providing coaching by nonmedical influencers to HCPs focused on social media strategies that generate impact (Hernandez et al. 2021).

5.4 Social Media as a Health Support Tool

Social media has also been reported to have positive influences in individuals dealing with other life threatening infections and or health challenges. The availability of social networks through social media has connected people all around the globe. This connectivity has allowed for the spread of information and creation of global online support groups where individuals share all kinds of information for peer support. Such groups also include those sharing health experiences for example their experience with certain disease or illness, including treatment and what has and or not worked for them. In mental health, peer support is seen as a promising strategy for recovery and overcoming limitations to illness (Mead et al. 2001; Naslund et al. 2014; Solomon 2004).

While previously peer support groups could have been done in person, online connectivity has widened the area of connectivity.

5.5 Conclusion

The importance of health information cannot be underrated. Currently, information is readily available through mainstream media and various social media platforms, it is important that accurate health information is disseminated. Misinformation has contributed to negative outcomes including vaccine hesitancy, poor adherence to health information, and substandard public health outcomes. By regulating information being distributed on social media and other media platforms and ensuring that accurate information is readily available to the public could assist in combating misinformation. By engaging healthcare professionals in dissemination of health information on various media platforms and setting up strategies to combat misinformation at all levels could assist in ensuring that the right information if disseminated. Achieving SDG3 and SDG3d in this era of misinformation requires better education (SDG4), strong and resilient infrastructures (SDG9) as well as political stability (peace). By improving healthcare systems and infrastructure in LMIC and ensuring peace and stability in the region could pave way for better health outcomes.

References

A Timeline of the Wakefield Retraction (2010) Nat Med 16:248. https://doi.org/10.1038/nm0310-248b

Abu-Odah H, Su J, Musa SS (2022) Unwillingness or reluctance of Palestinians to get the COVID-19 vaccine: the reasons behind it and how to persuade them. Int J Infect Dis 119:53. https://doi.org/10.1016/J.IJID.2022.03.039

Afzal SS, Qamar MA, Dhillon R, Bhura M, Khan MH, Suriya Q et al (2022) Budding medical professionals and COVID-19: the impact of COVID-19 on mental health and medical students. J Ayub Med Coll Abbottabad 34(3):483–488. https://doi.org/10.55519/JAMC-03-9572

American Psycholgical Association: APA Working Group on Health Disparities in Boys and Men (2018) Health Disparities in Racial/Ethnic and Sexual Minority Boys and Men, vol 2. https://www.apa.org/pi/health-equity/resources/race-sexuality-men

Amoah PA (2018) Social participation, health literacy, and health and well-being: a cross-sectional study in Ghana. SSM – Popul Health 4:263–270. https://doi.org/10.1016/j.ssmph.2018.02.005

Arcia A, Suero-Tejeda N, Spiegel-Gotsch N, Luchsinger JA, Mittelman M, Bakken S (2019) Helping hispanic family caregivers of persons with dementia "get the Picture" about health status through tailored Infographics. Gerontologist 59(5):e479–e489. https://doi.org/10.1093/geront/gnz085

Aylward B, Tangermann R (2011) The global polio eradication initiative: lessons learned and prospects for success. Vaccine 29(SUPPL. 4):D80–D85. https://doi.org/10.1016/j.vaccine.2011.10.005

Banerjee D, Kosagisharaf JR, Sathyanarayana Rao TS (2021) 'The dual pandemic' of suicide and COVID-19: a biopsychosocial narrative of risks and prevention. Psychiatry Res 295:113577. https://doi.org/10.1016/J.PSYCHRES.2020.113577

Bora K, Das D, Barman B, Borah P (2018) Are internet videos useful sources of information during global public health emergencies? A case study of YouTube videos during the 2015–16 Zika virus pandemic. Pathog Glob Health 112(6):320. https://doi.org/10.1080/20477724.2018.1507784

Borges do Nascimento IJ, Pizarro AB, Almeida JM, Azzopardi-Muscat N, Gonçalves MA, Björklund M, Novillo-Ortiz D (2022) Infodemics and health misinformation: a systematic review of reviews. Bull World Health Organ 100(9):544–561. https://doi.org/10.2471/BLT.21.287654

Boulware LE, Cooper LA, Ratner LE, LaVeist TA, Powe NR (2003) Race and trust in the health care system. Public Health Rep 118(4):358. https://doi.org/10.1016/S0033-3549(04)50262-5

Bresolin LB (1999) Health literacy: report of the council on scientific affairs. JAMA 281(6):552–557. https://doi.org/10.1001/jama.281.6.552

Budhathoki SS, Pokharel PK, Good S, Limbu S, Bhattachan M, Osborne RH (2017) The potential of health literacy to address the health related UN sustainable development goal 3 (SDG3) in Nepal: a rapid review. BMC Health Serv Res 17(1):1–13. https://doi.org/10.1186/s12913-017-2183-6

Caballero-Anthony M (2021) COVID-19 in Southeast Asia: regional pandemic preparedness matters. Nanyang Technological University. Order from Chaos [BLOG]. http://libaccess.mcmaster.ca/login?url=https://www.proquest.com/blogs-podcasts-websites/covid-19-southeast-asia-regional-pandemic/docview/2477799720/se-2?accountid=12347%0Ahttp://sfx.scholarsportal.info/mcmaster?url_ver=Z39.88-2004&rft_val_fmt=info:ofi/fm

Chandra R (2022) Challenges and vision in educating the global south. Int J Res Anal Rev 9(3):277–280

Chikovore J, Hart G, Kumwenda M, Chipungu GA, Corbett L (2015) "For a mere cough, men must just chew Conjex, gain strength, and continue working": the provider construction and tuberculosis care-seeking implications in Blantyre, Malawi. Glob Health Action 8(1):26292. https://doi.org/10.3402/gha.v8.26292

Cohen CR, Steinfeld RL, Weke E, Bukusi EA, Hatcher AM, Shiboski S et al (2015) Shamba Maisha: pilot agricultural intervention for food security and HIV health outcomes in Kenya: design, methods, baseline results and process evaluation of a cluster-randomized controlled trial. Springerplus 4(1):122. https://doi.org/10.1186/s40064-015-0886-x

DeWalt DA, Berkman ND, Sheridan S, Lohr KN, Pignone MP (2004) Literacy and health outcomes. J Gen Intern Med 19(12):1228–1239. https://doi.org/10.1111/j.1525-1497.2004.40153.x

Dionne KY, Turkmen FF (2020) The politics of pandemic Othering: putting COVID-19 in global and historical context. Int Organ 74(S1):E213–E230. https://doi.org/10.1017/S0020818320000405

Dredze M, Broniatowski DA, Smith MC, Hilyard KM (2016) Understanding vaccine refusal: why we need social media now. Am J Prev Med 50(4):550–552. https://doi.org/10.1016/j.amepre.2015.10.002

Dyer C (2010) Lancet retracts Wakefield's MMR paper. BMJ 340:696. https://doi.org/10.1136/BMJ.C696

Fagerlin A, Valley TS, Scherer AM, Knaus M, Das E, Zikmund-Fisher BJ (2017) Communicating infectious disease prevalence through graphics: results from an international survey. Vaccine 35(32):4041–4047. https://doi.org/10.1016/J.VACCINE.2017.05.048

Farooq S, Tunmore J, Wajid Ali M, Ayub M (2021) Suicide, self-harm and suicidal ideation during COVID-19: a systematic review. Psychiatry Res 306:114228. https://doi.org/10.1016/j.psychres.2021.114228

Feuerstein M (1999) Media literacy in support of critical thinking. J Educ Media 24(1):43–54. https://doi.org/10.1080/1358165990240104

Gibbs S (1996) Skin disease and socioeconomic conditions in rural Africa: Tanzania. Int J Dermatol 35(9):633–639. https://doi.org/10.1111/j.1365-4362.1996.tb03687.x

Gibbs A, Willan S, Misselhorn A, Mangoma J (2012) Combined structural interventions for gender equality and livelihood security: a critical review of the evidence from southern and eastern Africa and the implications for young people. J Int AIDS Soc 15(1):1–10. https://doi.org/10.1186/1758-2652-15-1

Godlee F, Smith J, Marcovitch H (2011) Wakefield's article linking MMR vaccine and autism was fraudulent. BMJ 342:d1678. https://doi.org/10.1136/BMJ.D1678

Hernandez RG, Hagen L, Walker K, O'Leary H, Lengacher C (2021) The COVID-19 vaccine social media *infodemic*: healthcare providers' missed dose in addressing misinformation and vaccine hesitancy. Hum Vaccin Immunother 17(9):2962–2964. https://doi.org/10.1080/21645515.2021.1912551

Iglesias Martínez E, Roces García J, Jiménez Arberas E, Llosa JA (2022) Difference between impacts of COVID-19 on women and Men's psychological, social, vulnerable work situations, and economic well-being. Int J Environ Res Public Health 19(14):8849. https://doi.org/10.3390/ijerph19148849

Irwan AM, Kato M, Kitaoka K, Kido T, Taniguchi Y, Shogenji M (2016) Self-care practices and health-seeking behavior among older persons in a developing country: theories-based research. Int J Nurs Sci 3(1):11–23. https://doi.org/10.1016/j.ijnss.2016.02.010

Jaiswal J, Halkitis PN (2019) Towards a more inclusive and dynamic understanding of Medical mistrust informed by science. Behav Med 45(2):79. https://doi.org/10.1080/08964289.2019.1619511

Kabir M, Afzal MS (2016) Epidemiology of polio virus infection in Pakistan and possible risk factors for its transmission. Asian Pac J Trop Med 9(11):1044–1047. https://doi.org/10.1016/j.apjtm.2016.09.006

The Lancet Psychiatry (2021) COVID-19 and suicide. Lancet Psychiatry 8:451. https://doi.org/10.1016/S2215-0366(21)00164-4

Levin-Zamir D, Lemish D, Gofin R (2011) Media health literacy (MHL): development and measurement of the concept among adolescents. Health Educ Res 26(2):323–335. https://doi.org/10.1093/HER/CYR007

Mack N, Odhiambo J, Wong CM, Agot K (2014) Barriers and facilitators to pre-exposure prophylaxis (PrEP) eligibility screening and ongoing HIV testing among target populations in Bondo and Rarieda, Kenya: results of a consultation with community stakeholders. BMC Health Serv Res 14(1):231. https://doi.org/10.1186/1472-6963-14-231

Marar SD, Al-Madaney MM, Almousawi FH (2019) Health information on social media. Saudi Med J 40(12):1294–1298. https://doi.org/10.15537/smj.2019.12.24682

Mead S, Hilton D, Curtis L (2001) Peer support: a theoretical perspective. Psychiatr Rehabil J 25(2):134–141. https://doi.org/10.1037/h0095032

Meites E, Gorbach PM, Gratzer B, Panicker G, Steinau M, Collins T et al (2016) Monitoring for human papillomavirus vaccine impact among gay, bisexual, and other men who have sex with men-United States, 2012–2014. J Infect Dis 214(5):689–696. https://doi.org/10.1093/infdis/jiw232

Mousley E, Deribea K, Tamiru A, Tomczyk S, Hanlon C, Davey G (2014) Mental distress and podoconiosis in northern Ethiopia: a comparative cross-sectional study. Int Health 7(1):16–25. https://doi.org/10.1093/inthealth/ihu043

Mukherjee S (2020) Disparities, desperation, and divisiveness: coping with COVID-19 in India. Psychol Trauma Theory Res Pract Policy 12(6):582. https://doi.org/10.1037/TRA0000682

Naslund JA, Grande SW, Aschbrenner KA, Elwyn G (2014) Naturally occurring peer support through social media: the experiences of individuals with severe mental illness using YouTube. PLoS One 9(10):e110171. https://doi.org/10.1371/journal.pone.0110171

Nematollahi S, Minter DJ, Hamaguchi R (2020) Ryoko Hamaguchi, medical student picture of a pandemic: visual aids in the COVID-19 crisis A validated COVID-19 infographic. Resid Physician J Public Health 42(3):483–485. https://doi.org/10.1093/pubmed/fdaa080

Norman CD, Skinner HA (2006) eHealth literacy: essential skills for consumer health in a networked world. J Med Internet Res 8:e9. https://doi.org/10.2196/jmir.8.2.e9

Nuwematsiko R, Nabiryo M, Bomboka JB, Nalinya S, Musoke D, Okello D, Wanyenze RK (2022) Unintended socio-economic and health consequences of COVID-19 among slum dwellers in Kampala, Uganda. BMC Public Health 22(1):1–13. https://doi.org/10.1186/S12889-021-12453-6/FIGURES/5

Omer SB (2020) The discredited doctor hailed by the anti-vaccine movement. Nature 586(7831):668–669. https://doi.org/10.1038/d41586-020-02989-9

Popoola BO (2019) Involving libraries in improving health literacy to achieve sustainable development Goal-3 in developing economies: a literature review. Health Inf Libr J 36(2):111–120. https://doi.org/10.1111/HIR.12255

Reny TT, Barreto MA (2022) Xenophobia in the time of pandemic: othering, anti-Asian attitudes, and COVID-19. Polit Groups Identities 10(2):209–232. https://doi.org/10.1080/21565503.2020.1769693

Schreiweis B, Pobiruchin M, Strotbaum V, Suleder J, Wiesner M, Bergh B (2019) Barriers and facilitators to the implementation of eHealth services: systematic literature analysis. J Med Internet Res 21(11):e14197. https://doi.org/10.2196/14197

Selvarajah S, Maioli SC, Deivanayagam TA, De P, Sato M, Devakumar D et al (2022) Racism, xenophobia, discrimination, and health 2 racism, xenophobia, and discrimination: mapping pathways to health outcomes. Lancet 400:2109–2124. https://doi.org/10.1016/S0140-6736(22)02484-9

Sepúlveda D, Mendoza Horvitz M, Joiko S, Ortiz Ruiz F (2022) Education and the production of inequalities across the global south and north. J Sociol 58(3):273–284. https://doi.org/10.1177/14407833211060059

Shahid R, Shoker M, Chu LM, Frehlick R, Ward H, Pahwa P (2022) Impact of low health literacy on patients' health outcomes: a multicenter cohort study. BMC Health Serv Res 22(1):1148. https://doi.org/10.1186/s12913-022-08527-9

Shannon G, Morgan R, Zeinali Z, Brady L, Thereza Couto M, Devakumar D et al (2022) Racism, xenophobia, discrimination, and health 3 intersectional insights into racism and health: not just a question of identity. Lancet 400:2125–2136. https://doi.org/10.1016/S0140-6736(22)02304-2

Smailhodzic E, Hooijsma W, Boonstra A, Langley DJ (2016) Social media use in healthcare: a systematic review of effects on patients and on their relationship with healthcare professionals. BMC Health Serv Res 16(1):442. https://doi.org/10.1186/s12913-016-1691-0

Solomon P (2004) Peer support/peer provided services underlying processes, benefits, and critical ingredients. Psychiatr Rehabil J 27(4):392–401. https://doi.org/10.2975/27.2004.392.401

Sørensen K, Van Den Broucke S, Fullam J, Doyle G, Pelikan J, Slonska Z, Brand H (2012) Health literacy and public health: a systematic review and integration of definitions and models. BMC Public Health 12(1):80. https://doi.org/10.1186/1471-2458-12-80

Sotgiu G, Dobler CC (2020) Social stigma in the time of coronavirus disease 2019. Eur Respir J 56(2):2002461. https://doi.org/10.1183/13993003.02461-2020

SteelFisher GK, Blendon RJ, Guirguis S, Brulé A, Lasala-Blanco N, Coleman M et al (2015) Threats to polio eradication in high-conflict areas in Pakistan and Nigeria: a polling study of caregivers of children younger than 5 years. Lancet Infect Dis 15(10):1183–1192. https://doi.org/10.1016/S1473-3099(15)00178-4

Stormacq C, Wosinski J, Boillat E, Van den Broucke S (2020) Effects of health literacy interventions on health-related outcomes in socioeconomically disadvantaged adults living in the community: a systematic review. JBI Evid Synth 18(7):1389–1469. https://doi.org/10.11124/JBISRIR-D-18-00023

Suarez-Lledo V, Alvarez-Galvez J (2021) Prevalence of health misinformation on social media: systematic review. J Med Internet Res 23(1):e17187. https://doi.org/10.2196/17187

Sylvia Chou W-Y, Gaysynsky A, Cappella JN (2020) Where we go from here: health misinformation on social media. Am J Public Health 110(S3):S273–S275. https://doi.org/10.2105/AJPH.2020.305905

Tasnim S, Hossain M, Mazumder H (2020) Impact of rumors and misinformation on COVID-19 in social media. J Prev Med Public Health 53:171–174. https://doi.org/10.3961/JPMPH.20.094

Thorburn S, Kue J, Keon KL, Lo P (2012) Medical mistrust and discrimination in health care: a qualitative study of Hmong women and men. J Community Health 37(4):822. https://doi.org/10.1007/S10900-011-9516-X

Traboco L, Pandian H, Nikiphorou E, Gupta L (2022) Designing Infographics: visual representations for enhancing education, communication, and scientific research. J Korean Med Sci 37(27):e214. https://doi.org/10.3346/jkms.2022.37.e214

Vall-Roqué H, Andrés A, Saldaña C (2021) The impact of COVID-19 lockdown on social network sites use, body image disturbances and self-esteem among adolescent and young women. Prog Neuro-Psychopharmacol Biol Psychiatry 110:110293. https://doi.org/10.1016/J.PNPBP.2021.110293

Van Hecke O, Lee JJ, Butler CC, Moore M, Tonkin-Crine S (2020) Using evidence-based infographics to increase parents' understanding about antibiotic use and antibiotic resistance: a proof-of-concept study. JAC-Antimicrob Resist 2(4):dlaa102. https://doi.org/10.1093/JACAMR/DLAA102

Vlassoff C (2007) Gender differences in determinants and consequences of health and illness. J Health Popul Nutr 25(1):47–61

6 Equitable Health Response: Lessons from COVID-19 and Monkeypox

Abstract

The COVID-19 pandemic has affected populations worldwide with LMIC being greatly affected, not only by COVID-19 but the disruption of various disease management programs which have derailed the progress that has been achieved in the past decades. What lessons have been learned in managing a pandemic within another pandemic? What is the way forward in public health response and management in LMIC? How can LMIC sustain public health response and management in the wake of other challenges such as the COVID-19 pandemic?

It is essential that the health systems in LMIC are strengthened and improved to meet the global standards and provide primary healthcare services for their populations. The current state of healthcare systems needs to be improved before any discussion of sustainability. Sustaining health systems in LMIC in their current state will just further increase the healthcare gap.

Keywords

Equitable health response · Low and middle income countries · COVID-19 · Monkeypox · Access · Sustainability · Discrimination · Connectivity · Internet · Low and middle income countries · Inequalities · Equity · Vaccine distribution · Pandemic response

6.1 Managing a Pandemic Within Other Pandemics

The COVID-19 pandemic is not and will not be the last pandemic to be encountered. The current trend in emerging and re-emerging diseases depicts a higher likelihood of another event of similar magnitude if not greater. In the last decade there have been an increase in incidence of emerging and re-emerging diseases globally and LMIC carry the burden (Fig. 6.1). One of the most recent emerging diseases with devastating consequences has been the COVID-19 (SARS-CoV2) which started as an outbreak in China but spread globally resulting in a distressing pandemic.

As of February 2023, 756,581,850 confirmed cases of COVID-19, including 6,844,267 deaths, were reported to WHO (https://covid19.who.int/). The number of cases from the highest to the least cases recorded, were in Europe, followed by the Americas, Western Pacific, Southeast Asia, Eastern Mediterranean and Africa with the least number of cases (https://covid19.who.int/).

Rapid response was required to combat the growing pandemic, these included therapeutics such as drugs and vaccines that could assist in reducing the effect of the virus. Various

F. A. Mphande, *Sustainable Health in Low and Middle Income Countries*, Sustainable Development Goals Series, https://doi.org/10.1007/978-981-99-4254-1_6

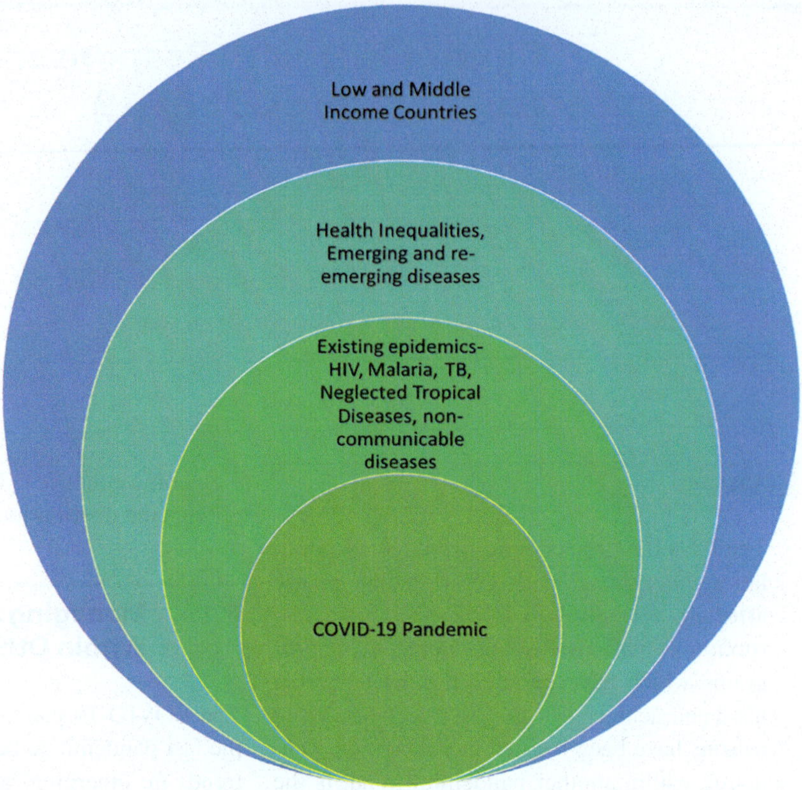

Fig. 6.1 Health Burden and LMIC, LMIC carry a huge health burden, these include the COVID-19 pandemic, existing epidemics, neglected tropical diseases, emerging and re-emerging diseases, health non-communicable diseases and health inequalities which negatively affect equitable health

companies worked to produce vaccines which were made available to masses worldwide. The vaccine acquisition and distribution had its own challenges. Hoarding of vaccines by high income countries was one of the concerning responses with LMIC struggling to access these lifesaving tools. Despite the skewed distribution of vaccines, 13,008,560,983 vaccine doses have been administered globally (https://covid19.who.int/). Through COVAX a lot of LMIC were able to access the vaccines although the doses were minimal compared to the countries' population (Fig. 6.2) (Tagoe et al. 2021).

The challenges observed in the implementation of vaccine distribution during the COVID-19 pandemic not only impedes achieving SDG3.3 (By 2030, end the epidemics of AIDS, tuberculosis, malaria and neglected tropical diseases and combat hepatitis, water-borne diseases and other communicable diseases).

Another outbreak that has affected several countries is that of Monkeypox. Identified in 1970 from a 9 month old baby in the Democratic Republic of Congo (DRC), cases have been on the rise since then with frequent outbreaks in West Africa. In 2003, the first monkeypox outbreak outside Africa was reported in the United States of America. Previously, outbreaks outside the African continent have been reported in travellers to the UK (2018, 2019 and 2021), and Singapore (2019 and 2021). The recent outbreak of Monkeypox with the most recorded cases and with no links to travellers was reported in Europe, USA and Latin America.

6.2 Lessons from the Pandemic

To better understand the effects of the pandemic, it is important to familiarize with the various determinants of health. These include genetics, behaviour, environmental and physical influences, medical care, and social factors. Genetic makeup can affect whether one is susceptible to

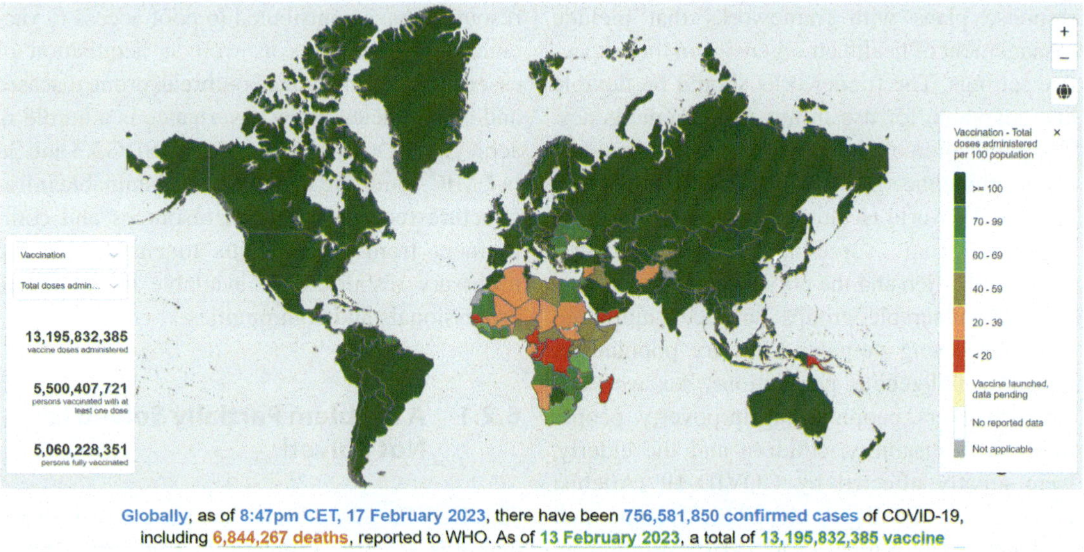

Fig. 6.2 Global COVID-19 vaccination coverage as of February 2023. Source WHO Coronavirus (COVID-19) Dashboard https://covid19.who.int/?adgroupsurvey={adgroupsurvey}&gclid=CjwKCAiA0cyfBhBREiwAAtStHIWdisDIKIWizZ5MVME9ENSOoK_4gVjlUBom8gd9hspgnJMV-aH0WRoC1IkQAvD_BwEhttps://covid19.who.int/?adgroupsurvey={adgroupsurvey}&gclid=CjwKCAiA0cyfBhBREiwAAtStHIWdisDIKIWizZ5MVME9ENSOoK_4gVjlUBom8gd9hspgnJMV-aH0WRoC1IkQAvD_BwE

certain diseases or are at risk of developing certain diseases and conditions. Human behaviour as well as the environment can increase the risk of disease through activities that could expose and or increase the risk of disease; while living in a disease endemic area, increases the likelihood of exposure to a certain disease. Availability of physical infrastructure, and dependable healthcare services as well as an environment that can ensure health and wellbeing of a population is thus essential. Developing resilient and sustainable infrastructure in LMIC including those that could assist in developing therapies and vaccines for use in LMIC is a goal that these countries must strive to achieve. Thus, SDG3b (Support the research and development of vaccines and medicines for the communicable and non-communicable diseases that primarily affect developing countries, provide access to affordable essential medicines and vaccines, in accordance with the Doha Declaration on the TRIPS Agreement and Public Health, which affirms the right of developing countries to use to the full the provisions in the Agreement on Trade-Related Aspects of Intellectual Property Rights regarding flexibilities to protect public health, and, in particular, provide access to medicines for all) should be a top priority for LMIC. The opening of advanced laboratories in the WHO African region will support research capabilities within the region and could help reduce the gaps in access to drugs and vaccines which could be produced locally in the coming years.

The COVID-19 pandemic has affected various areas of health and has exposed gaps in the healthcare system as well as global health response. The gaps include limited availability of data which could be attributed to (but not limited) poor and or lack of surveillance, poor data management, limited capacity to analyze available data as well as governance (Jayatilleke 2020; Trolle et al. 2023). As such enhancing capacity in data acquisition and management could assist in strengthening the likelihood of attaining SDG3d. While some infectious diseases such as malaria, TB, Hepatitis, may have thriving systems in place, emerging and re-emerging diseases may not have systems in place making management of such diseases difficult. It thus crucial that LMIC should invest in emergency preparedness and response plans with frameworks that include management of health emergencies in the respective settings. The frameworks should be flexible and adaptable for use in any health emergency. The publication of the WHO priority list could be useful guideline (Simpson et al. 2020; Sorvillo et al. 2020; World Health Organization 2020).

Apart from surveillance, discrimination between the rich and the poor nations, the minority, and vulnerable groups has been amplified with concerning outlook. Minority populations including indigenous populations, sex workers, casual workers, people living in poverty, people living with disability, children and the elderly, were greatly affected by COVID-19 (Adebisi et al. 2020; Cahill et al. 2020; Dionne and Turkmen 2020; Knight et al. 2020; Wang et al. 2021). In other parts of the world, isolated populations, such as indigenous populations had difficulties in accessing care and treatment during the pandemic. It is difficult to estimate the numbers due to limited surveillance in such isolated populations (Agoramoorthy and Hsu 2022; Argoty-Pantoja et al. 2021; Humeyestewa et al. 2021; Huyser et al. 2022; Machado et al. 2022; Muradian et al. 2013; Waitoki and McLachlan 2022; Wichmann and Wichmann 2022). Thus, establishing resilient healthcare facilities and systems for indigenous populations is essential. Implementation of such systems would require commitment from respective governments to ensure that the required infrastructure is in place. These would help with access to treatment, vaccines, and other health services to tackle health emergencies in remote and difficult to reach populations. This would be a step towards achieving SDG3d in vulnerable populations and LMIC.

Physical infrastructure and dependable healthcare system requires financial resources, to build, manage, maintain, and sustain. With LMIC financial challenges are a major part of the status of health systems today. There are other factors including governance which have hindered development of sustainable healthcare systems. Political upheavals, corruption and poor resource management has seen most LMIC struggle to build and manage sustainable healthcare systems to the detriment of the society. Limited financial resources have contributed to poor access to vaccines and treatments as well as acquisition of essential tools to manage outbreak prone diseases and emergencies. Poor governance is a hurdle in achieving SDG3, and specifically SDG3.3 and 3d in LMIC. Building resilient and sustainable infrastructure requires financial resources and commitment from governments to ensure that the necessary resources are available to healthcare professionals and communities.

6.2.1 A Problem Partially Solved Is Not Solved

The COVID-19 outbreak has shown that a pandemic is a global problem. Despite some countries having better healthcare systems than others, advanced equipment, and technologies, the pandemic will still spread to all regions. As such one major lesson from this pandemic is that the management of a pandemic should be universal and not discriminatory. Any form of discrimination often results in negative responses that affect the management and control of the disease in question. Discrimination was observed in vaccine acquisition and distribution, targeted closures of borders for certain populations, marginalization of indigenous and vulnerable populations in treatment and control (Eccleston-Turner and Upton 2021; Riaz et al. 2021; Tagoe et al. 2021). Racial discrimination and its impact on health cannot be overlooked in the modern-day world and are important determinants of health (Abubakar et al. 2022; Dionne and Turkmen 2020.; Selvarajah et al. 2022). Racism, xenophobia, and discrimination can be in different forms, including microaggression, interpersonal and violence which affects health in many different ways. Structural racism has affected many with minorities affecting both their mental and physical health (Paradies et al. 2015; Selvarajah et al. 2022). The COVID-19 severely affected minority populations with disparities observed within the countries in high income countries as well as between high income and LMIC (Mukherjee 2020).

To date racial discrimination has been incorporated in health policies, health systems, and treatment rooms as part of standard practice through systemic oppressions and interpersonal aggressions most of which are accepted as a norm. There have been instances where certain populations have been denied appropriate treatment due to their race while other populations are given all kinds of medications and treatments (Devakumar et al. 2022; Jaiswal and Halkitis 2019; Shannon et al. 2022). One will be put through certain diagnostic tests or denied certain diagnostic tests because of their race, skin colour or religion. History has evidence of how racism and discrimination has affected health (Devakumar et al. 2022; Jaiswal and Halkitis 2019). COVID 19 response was an eye opener on how these discriminatory ideologies can play a huge role in disease response and management. Vaccine distribution was one such area. While vaccines were widely distributed in high income countries, with >100 doses administered per 100 population (Fig. 6.2), most LMIC had lower doses administered. This type of disparity could lead to COVID-19 breeding in pockets with less vaccination coverage. These pockets become breeding grounds for mutations creating variants some of which could be more virulent and outbreak prone (Haque and Pant 2022). There is a high possibility that these variants may not recognized by the current vaccines and therapies (Fernandes et al. 2022; Haque and Pant 2022). The mutant variants could spread within the community. These communities would then act as reservoirs and giving opportunity to creation of more variants as well as spreading the disease to vaccinated communities who may not have immunity against the new variants. Thus, while some countries try to find reasons to support discriminatory ideologies and acts, addressing the root cause of these ideologies could be one step in creating health equity and equality (Devakumar et al. 2022; Jaiswal and Halkitis 2019; Mukherjee 2020; Selvarajah et al. 2022). Discriminatory ideologies negatively affect chances of meeting various SDG including SDG3.3 and 3d. If a health emergency is viewed under the lens of "them" and "us", the chances of attaining sustainable development goals in LMIC are slim.

6.2.2 Limited Access to Resources

A rather intriguing pattern of response to infectious disease has been observed in the past decades with some of the major outbreaks of Ebola, COVID-19 and the recent Monkeypox virus. While outbreaks maybe ongoing in LMIC as long as they remain in those regions, it may not catch as much media attention until it becomes a threat to high income countries. This type of response has resulted in limited action in terms of resources, including investment in treatment in the affected populations. The response to COVID-19 pandemic and the Monkeypox outbreak in Europe is evidence to such a response. While various treatments and vaccines were made available in high income communities, LMIC remain behind on accessing such crucial resources. The events during the COVID-19 pandemic and the current Monkeypox outbreak have shown a growing divide in global health response between HIC and LMIC as well as the rich and the poor (within the same country). According to a commission report published in Lancet, multiple failures were identified. At global level, timely notification, as well as response to slow down the spread of the virus, equitable distribution of resources, and lack of coordination among countries were identified. At national level countries were slow to adopt and implement strategies that could slow down the pandemic with lack of finances and equipment further contributing to inadequate response in LMIC. Health systems and inadequate data also affected response at national level (Sachs et al. 2022). Unlike smallpox which was eradicated with joint efforts to end the dreadful disease, the response to the pandemic was marred by intellectual property rights, technology transfer, international financing, which affected vaccine production and distribution in LMIC (Sachs et al. 2022). The current multilateral efforts display a rather grim picture with considerable effects on attaining SDG.

Media coverage was crucial during the pandemic as it kept the public informed on the pandemic. While the information helped track the pandemic globally, it created fear and anxiety in some circles affecting individuals' mental health (Anwar et al. 2020). Lack of homogeneity and little or no consideration of other cultural values have contributed to biased reporting at times with remarks that are rather insensitive to other cultures. This type of reporting often contributed to loss of interest and lack of trust to the information provided by the media as well as associated governments. As such even if such media outlets were to promote relevant health information, populations that feel unrepresented would be less likely to accept or follow it (Lee and You 2022; Ng and Tan 2022). Limited access to information created gaps with resulting in lack of crucial information in communities paving way to misinformation and infodemics in the affected communities (Matta 2020).

Information dissemination is crucial in achieving SDG3.3. Public health campaigns against infectious diseases such as malaria, TB, diarrhoeal and other outbreak prone infections have contributed to improved health outcomes in the populations. Negative information by media would negatively affect the progress made with these public health campaigns.

Information dissemination in the twenty-first century is fast with various modes of communication. While mainstream media, TV, radio and print media (newspapers, magazines, etc.) still exist, social media platforms have taken over with far more reach than the former. Social media platforms can be easily accessed on portable gadgets such as mobile phones and tablets, making them readily available to the public anywhere in the world as long as there is internet connectivity. This has made social media platforms as a go to resource if one needs to reach a wider audience. This mode of communication has resulted in success of health campaigns but also failure of others (Abu-Odah et al. 2022; Bora et al. 2018; Borges do Nascimento et al. 2022; Chapman et al. 2018; Nematollahi et al. 2020; Rupprecht and Burgess 2015). Poor regulation of material posted on social media has resulted in the spread of distorted facts, which often have steered public health campaigns in the wrong direction. Limited access to internet has thus deprived communities of crucial health information that could be accessible online (Levin-Zamir and Bertschi 2018; Neubeck et al. 2009). Subsequently, distorted facts have been accepted as truth as populations with lack of access to information may not be able to distinguish what is true and or false (Suarez-Lledo and Alvarez-Galvez 2021; Sylvia Chou et al. 2020). Ensuring that populations in LMIC have access to factual information, could have benefit such population in achieving SDG3.3.

6.2.3 Need for Self-Sustainability and Change of Mindset

"Vaccine nationalism" is one new terminology that gained popularity during the COVID-19 pandemic (Eccleston-Turner and Upton 2021; Kretchmer 2021; Riaz et al. 2021). As high-income countries hoarded vaccine doses, LMIC were left out. This led to limited availability for the vaccine to over 6 billion people living in LMIC. This gesture prompted many countries to carefully consider how to manage future pandemics. On the African continent, improvements were suggested to advance infrastructure and enhance capacities both at country and regional level (Nkengasong et al. 2017). With less than 1% of vaccines produced on the African continent, there is need to encourage vaccine production on the continent which is battling a wide range of vaccine preventable diseases. Lack of financial resources and sustainable production of equipment and infrastructure to produce these vaccines and other therapeutics remains a challenge (Adepoju 2022). Regional integrated networks through Africa CDC could assist in improving laboratory capacity, surveillance and data sharing within the region (Adeniran et al. 2022). In countries such as Uganda which has experienced both COVID-19 and Ebola disease outbreaks, the country was able to apply the knowledge gained during these previous outbreaks to combat monkeypox outbreak (Tusabe et al. 2022). Similarly, Nigeria expanded its laboratory capacity by expanding their laboratory network from 12 to 77 (Tusabe et al. 2022). At the regional level, vaccine production facilities were established in Algeria, Senegal, Egypt, Morocco, Rwanda, and South Africa with Twelve vaccine production facilities expected to produce a various types of COVID-19 vaccines. The implementation of the Partnership for African Vaccine Manufacturer (PAVM) framework for Action 2022, has facilitated technology transfer on mRNA vaccines to the continent. Despite the laboratory capabilities in some countries, generally, most countries in Africa struggle with limited diagnostic capacity and budgetary allocations to enhance diagnostics, training of human resources, and international accreditation (Tusabe et al. 2022). These developments would assist in fighting vaccine preventable diseases as well as establishing resilient infrastructure on the continent.

Apart from vaccines, various collaborative programmes have been launched that would assist implementing International Health Regulations (IHR) in the Africa region. Various training programmes such as field epidemiology training, field laboratory leadership programmes, public health informatics training programmes and adapting as well as recognizing community healthcare workers at national level was encouraged (Nkengasong et al. 2017). Community Healthcare workers form an essential workforce in primary healthcare in LMIC (Pallas et al. 2013). Community Healthcare workers (CHW), also known as village health workers in other countries, are individuals who work with and within the community providing basic health services and healthcare. Community Healthcare workers are often members of the communities they work with as such are trusted by the members of the community, thereby promoting compliance within the community (Lewin et al. 2010). Lack of professional certification, and often working on a volunteer basis without salary or minimal compensation, HCW assist in driving and achieving various disease prevention programmes and strengthen health systems in resource limited settings (Haines et al. 2007; Hermann et al. 2009). How can LMIC capitalize on this existing network of primary healthcare workers to achieve outbreak preparedness and response? A change in mindset is required and CHW need to be empowered and given proper training and certification (Lewin et al. 2010). Could the COVID-19 pandemic be a turning point for Africa and other LMIC?

Apart from COVID-19, the Monkeypox outbreak further highlights the existing gaps in access to therapies and vaccines in LMIC and in vulnerable populations (Yinka-Ogunleye et al. 2019). While diagnosis can be rapid in HICs, LMIC are limited not only in diagnostic tools but also infrastructure to carry out the diagnosis as well as vaccines and therapies that are required to prevent the spread of the disease. In Latin America region, access to improved therapies and vaccines is required to enhance preparedness and response (Rodriguez-Morales et al. 2022). Rodriguez-Morales et.al pointed several lessons learned from the COVID-19 pandemic and Monkeypox outbreak in Latin America. One of the challenges observed were failures in risk communication. The team therefore suggested strengthening epidemiological surveillance systems within the region and dissemination of adequate and reliable information. Syndemics, was another challenge, where countries are faced with more than one pandemic and have to manage these pandemics simultaneously. The COVID -19 pandemic affected the sustainability of existing disease control programmes such as TB, HIV, malaria, dengue, Chikungunya, Zika and neglected diseases. Countries must therefore find ways to manage syndemics sustainably. Prioritizing surveillance, early disease detection and close monitoring of potential zoonotic transmissions could assist in preparedness within the region (Rodriguez-Morales et al. 2022).

Sustained investment in various approaches including research, contact tracing, surveillance, and identification pf emerging and re-emerging diseases in LMIC is crucial. Currently, countries in LMIC are struggling to sustain their current health systems most of which have been weakened by the various ongoing pandemics including HIV/AIDS, malaria, TB and COVID-19.

Sustained investment requires good governance which is willing to manage and sustain disease control and prevention programs as well as encouraging multi sector approach in outbreak preparedness and response. Sustained support from government leaders, including policy makers and cross border cooperation could strengthen efforts to attain SDG3.3 and SDG3d in LMIC.

6.2.4 Stigma and Compliance

Stigma was observed during the COVID-19 pandemic and Monkeypox outbreak globally. With limited information of the disease at the onset of the outbreaks, knowledge gaps developed, and these were fed by misinformation which created fear, and anxiety within communities. The resulting stigma affected health seeking behaviour, compliance to vaccines and other health related prevention measures (Banerjee et al. 2021; Dionne and Turkmen 2020; Happi et al. 2022; Pillay et al. 2021; Tamang and Dorji 2022).

While pre-exposure prophylaxis (PrEP), antiviral therapies and vaccines are available for Monkeypox in HICs; these are not accessible to most LMIC including endemic countries. As such prevention and control of transmission of monkeypox remains crucial in limited resource settings (Karbalaei and Keikha 2022; Yinka-Ogunleye et al. 2019).

6.2.5 Need for Connectivity

Staying connected is an important aspect of society. Connections could be in form of social networks within the community, between organizations, having access to information through news and any form of media as well as having access to the information. Social networking has grown tremendously in the past decade and has become a powerful force for connecting people around the world. Social networking has been used to connect people for various causes including health related issues such as mental health. Use of social media for health purposes has been reported in LMIC for purposes of dissemination of health information, surveillance, as well as communication between healthcare providers and collaborators (Abiad et al. 2017; McCool et al. 2022; Wallis et al. 2017). Social media has also been used negatively to spread misinformation and cultivate negative responses to health prevention measures (Hagg et al. 2018). Why is connectivity important?

6.2.5.1 Access to Internet

Having a reliable internet source makes a huge difference in how people can access information as well as how much information they can have access to (Ekeland et al. 2010; Levin-Zamir and Bertschi 2018; Norman and Skinner 2006). Despite over 50% of the population in low-income countries having cellular subscription, less than 1% of the population has access to internet (https://data.worldbank.org/indicator/IT.NET.USER.ZS). This implies that 99% of the population in low-income countries may not have access to any health information, materials and resources provided online. Over 6 billion people live in LMIC and only 54% of the population uses internet (Fig. 6.3). Thus 46% of the population that does not use the internet could miss out on various information available online. The information may include health education, campaigns, publications, access to webinars, online courses, virtual conferences, workshops and much more. The lack of access to such valuable tools has further enhanced the knowledge gap between populations that have access to internet compared to those that do not have access (Abiad et al. 2017).

During the COVID-19 pandemic, lockdowns were crucial in limiting contact and preventing the spread of the virus, despite this, lockdowns had a great impact on mental health and social wellbeing of the populations involved. Lack of internet access meant that there was very limited interactions with the community and disruption of social connectivity (Sepúlveda-Loyola et al. 2020). While high income countries and some middle income countries (MIC) were able to shift activities online, most LMIC did not have the capacity to do so (Kroese et al. 2021). Despite the limited access to the internet, cellular

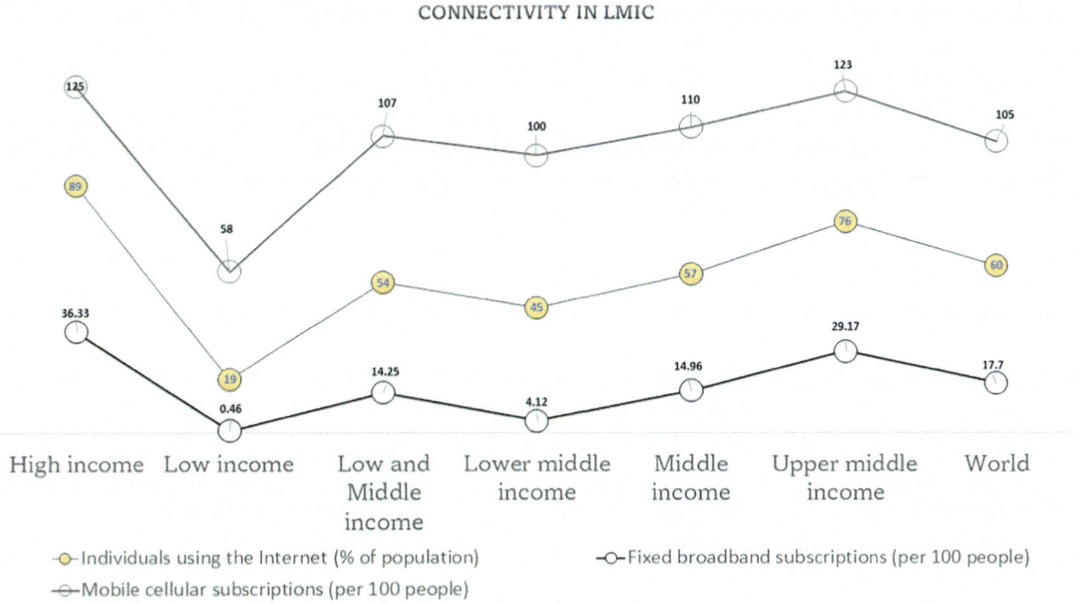

Fig. 6.3 Connectivity in LMIC, although mobile cellular subscriptions are high in LMIC the number of individuals using the internet is approximately 50% and less than 20% have fixed broadband subscriptions

subscriptions were much higher (Fig. 6.1). Thus, connectivity through mobile phones could prove much more effective in LMIC although further exploration is required. In Kosovo, online and phone psychological first aid services were used as an intervention during the COVID-19 pandemic. Tools such as psychoeducational videos and webinars were used to reach communities (Arenliu et al. 2020).

6.3 Equitable Health Response and Sustainability

Lessons from the recent Monkeypox and COVID -19 responses, as well as responses to previous outbreaks of global concern including Ebola, LMIC are still lagging behind, and equitable health response is far from being achieved. Progress that had been made in global health in the past decades for SDG3 have been greatly affected. Disruptions in healthcare services were reported in 92% of the countries, while universal health coverage came to a sudden halt. The global life expectancy and immunization coverage decreased while an increase in mental health conditions such as anxiety and depression and deaths from existing pandemic diseases such as TB and malaria were observed. In the year 2020 alone, 22.7 million children missed their basic vaccinations and the healthcare community lost over 100,000 healthcare workers (United Nations 2021).

There is need to consider how the SDG3 will be achieved in light with emerging and re-emerging diseases as well as public health emergencies of global concern. The COVID-19 pandemic has unearthed gaps that exist in healthcare systems and global health response systems that not only threaten universal health coverage, but also wellbeing for all at all ages. Disparities between the rich and the poor existing within LMIC and disparities between HIC, and LMIC further drive the challenges to equitable health and universal health coverage.

Nationalistic approaches to tackle the COVID-19 pandemic not only impede achieving SDG3 but weaken the fight against emerging and re-emerging infectious disease threats globally. As long as there is an existing reservoir in any corner of the world, and there are many potential susceptible hosts, it just a matter of time before another pandemic arises. The successful response to the COVID-19 pandemic is commendable but the gaps in the response system have to be addressed if SDG3 it to be achieved.

Sustainable development goal 3d states that "Strengthen the capacity of all countries, in particular developing countries, for early warning, risk reduction and management of national and global health risks". Efforts are being made in LMIC to strengthen capacities, and improve management of emerging and re-emerging disease threats (Chua et al. 2020; Lee et al. 2020; Nkengasong et al. 2017; Rodriguez-Morales et al. 2022).

Vaccine inequity, poor access to advanced diagnostic tool and therapies, limited access to information as well as systemic hurdles designed to limit certain populations from accessing valuable tools remain a challenge to equitable health and sustainable responses (Budhathoki et al. 2017; Burton et al. 2021; Domapielle et al. 2022; Eccleston-Turner and Upton 2021; Khankeh et al. 2021).

6.4 Conclusion

COVID-19 is not the last pandemic to occur, but a practical lesson on the challenges brought about by a highly infectious emerging disease capable of rapid to human transmission and fast mutation rates. Equitable health and response are still far from being achieved but countries are working hard to improve and scale up their responses and enhance their health systems and capacity. Governance remains a challenge, as this determines the implementation and success of interventions. Governance remains a major contributing factor in the sustainability of health systems in LMIC. It is essential that the health systems in LMIC are strengthened and improved to meet the global standards and provide primary healthcare services for their populations. The current state of healthcare systems needs to be improved before any discussion of sustainability. Sustaining health systems in LMIC in their current state will just further increase the healthcare gap. Equitable health responses are possible, but only if the gaps in healthcare preparedness and response are improved. It is a dilemma to talk about equitable health response when different settings experience the same problem are equipped with different tools. Those with better tools have a greater chance of success than those with limited tools.

References

Abiad A, Debuque-gonzales M, Sy AL, Abiad A, Debuque-gonzales M (2017) The role and impact of infrastructure in middle – income countries: anything special? ADB economics working paper series. The role and impact of infrastructure in middle-income, vol 518. University of the Philippines

Abubakar I, Gram L, Lasoye S, Achiume T, Becares L, Kaur Bola G et al (2022) Confronting the consequences of racism, xenophobia, and discrimination on health and health-care systems. Lancet 400:2137–2146. https://doi.org/10.1016/S0140-6736(22)01989-4

Abu-Odah H, Su J, Musa SS (2022) Unwillingness or reluctance of Palestinians to get the COVID-19 vaccine: the reasons behind it and how to persuade them. Int J Infect Dis 119:53. https://doi.org/10.1016/J.IJID.2022.03.039

Adebisi YA, Alaran AJ, Akinokun RT, Micheal AI, Ilesanmi EB, Lucero-Prisno DE (2020) Sex workers should not be forgotten in Africa's COVID-19 response. Am J Trop Med Hyg 103(5):1780. https://doi.org/10.4269/AJTMH.20-1045

Adeniran AA, Hamill LC, Selby R, Downs P (2022) COVID-19: A unique opportunity to improve laboratory capacity for neglected tropical diseases in sub-Saharan Africa. Front Trop Dis 3:984906. https://doi.org/10.3389/fitd.2022.984906

Adepoju P (2022) Africa CDC warns COVID-19 vaccine production could cease. Lancet 399(10336):1683. https://doi.org/10.1016/S0140-6736(22)00775-9

Agoramoorthy G, Hsu MJ (2022) COVID-19 and India's vulnerable indigenous populations. Am J Hum Biol 34(2):e23608. https://doi.org/10.1002/AJHB.23608

Anwar A, Malik M, Raees V, Anwar A (2020) Role of mass media and public health communications in the COVID-19 pandemic. Cureus 12(9):e10453. https://doi.org/10.7759/cureus.10453

Arenliu A, Uka F, Weine S (2020) Building online and telephone psychological FIRST aid services in A low resource setting during COVID-19: the CASE of Kosovo. Psychiatr Danub 32(3–4):570–576. https://doi.org/10.24869/psyd.2020.570

Argoty-Pantoja AD, Robles-Rivera K, Rivera-Paredez B, Salmerón J (2021) COVID-19 fatality in Mexico's indigenous populations. Public Health 193:69. https://doi.org/10.1016/J.PUHE.2021.01.023

Banerjee D, Kosagisharaf JR, Sathyanarayana Rao TS (2021) 'The dual pandemic' of suicide and COVID-19: a biopsychosocial narrative of risks and prevention. Psychiatry Res 295:113577. https://doi.org/10.1016/J.PSYCHRES.2020.113577

Bora K, Das D, Barman B, Borah P (2018) Are internet videos useful sources of information during global public health emergencies? A case study of YouTube videos during the 2015–16 Zika virus pandemic. Pathog Global Health 112(6):320. https://doi.org/10.1080/20477724.2018.1507784

Borges do Nascimento IJ, Pizarro AB, Almeida JM, Azzopardi-Muscat N, Gonçalves MA, Björklund M, Novillo-Ortiz D (2022) Infodemics and health misinformation: a systematic review of reviews. Bull World Health Organ 100(9):544–561. https://doi.org/10.2471/BLT.21.287654

Budhathoki SS, Pokharel PK, Good S, Limbu S, Bhattachan M, Osborne RH (2017) The potential of health literacy to address the health related UN sustainable development goal 3 (SDG3) in Nepal: a rapid review. BMC Health Serv Res 17(1):1–13. https://doi.org/10.1186/s12913-017-2183-6

Burton MJ, Ramke J, Patricia Marques AA, Bourne RR, Congdon N, Jones I et al (2021) The lancet global health commission the lancet global health commission on global eye health: vision beyond 2020. Lancet Glob Health 9(4):e489–e551. https://doi.org/10.1016/S2214-109X(20)30488-5

Cahill S, Grasso C, Keuroghlian A, Sciortino C, Mayer K (2020) Sexual and gender minority health in the COVID-19 pandemic: why data collection and combatting discrimination matter now more than ever. Am J Public Health 110(9):1360. https://doi.org/10.2105/AJPH.2020.305829

Chapman GE, Baylis M, Archer DC (2018) Survey of UK horse owners' knowledge of equine arboviruses and disease vectors. Vet Rec 183(5):159. https://doi.org/10.1136/vr.104521

Chua AQ, Tan MMJ, Verma M, Han EKL, Hsu LY, Cook AR et al (2020) Health system resilience in managing the COVID-19 pandemic: lessons from Singapore. BMJ Glob Health 5(9):e003317. https://doi.org/10.1136/bmjgh-2020-003317

Devakumar D, Selvarajah S, Abubakar I, Kim S-S, McKee M, Sabharwal NS et al (2022) Racism, xenophobia, discrimination, and the determination of health. Lancet 400(10368):2097–2108. https://doi.org/10.1016/S0140-6736(22)01972-9

Dionne KY, Turkmen FF (2020) The politics of pandemic Othering: putting COVID-19 in global and historical context. Cambridge University Press, pp 213–230. https://doi.org/10.1017/S0020818320000405

Domapielle MK, Sumankuuro J, Der Bebelleh F (2022) Revisiting the debate on health financing in low and middle-income countries: an integrative review of selected models. Int J Health Plann Manag 37(6):3061–3074. https://doi.org/10.1002/HPM.3566

Eccleston-Turner M, Upton H (2021) International collaboration to ensure equitable access to vaccines for COVID-19: the ACT-accelerator and the COVAX facility. Milbank Q 99(2):426–449. https://doi.org/10.1111/1468-0009.12503

Ekeland AG, Bowes A, Flottorp S (2010) Effectiveness of telemedicine: a systematic review of reviews. Int J Med Inform 79(11):736–771. https://doi.org/10.1016/j.ijmedinf.2010.08.006

Fernandes Q, Inchakalody VP, Merhi M, Mestiri S, Taib N, Moustafa Abo El-Ella D et al (2022) Emerging COVID-19 variants and their impact on SARS-CoV-2 diagnosis, therapeutics and vaccines. Ann Med 54(1):524–540. https://doi.org/10.1080/07853890.2022.2031274

Hagg E, Dahinten VS, Currie LM (2018) The emerging use of social media for health-related purposes in low and middle-income countries: a scoping review. Int J Med Inform 115:92–105. https://doi.org/10.1016/j.ijmedinf.2018.04.010

Haines A, Sanders D, Lehmann U, Rowe AK, Lawn JE, Jan S et al (2007) Achieving child survival goals: potential contribution of community health workers. Lancet 369(9579):2121–2131. https://doi.org/10.1016/S0140-6736(07)60325-0

Happi C, Adetifa I, Mbala P, Njouom R, Nakoune E, Happi A et al (2022) Urgent need for a non-discriminatory and non-stigmatizing nomenclature for monkeypox virus. PLoS Biol 20(8):e3001769. https://doi.org/10.1371/journal.pbio.3001769

Haque A, Pant AB (2022) Mitigating Covid-19 in the face of emerging virus variants, breakthrough infections and vaccine hesitancy. J Autoimmun 127:102792. https://doi.org/10.1016/j.jaut.2021.102792

Hermann K, Van Damme W, Pariyo GW, Schouten E, Assefa Y, Cirera A, Massavon W (2009) Community health workers for ART in sub-Saharan Africa: learning from experience – capitalizing on new opportunities. Hum Resour Health 7(1):31. https://doi.org/10.1186/1478-4491-7-31

Humeyestewa D, Burke RM, Kaur H, Vicenti D, Jenkins R, Yatabe G et al (2021) COVID-19 response by the Hopi tribe: impact of systems improvement during the first wave on the second wave of the pandemic. BMJ Glob Health 6(5):5150. https://doi.org/10.1136/BMJGH-2021-005150

Huyser KR, Yellow Horse AJ, Collins KA, Fischer J, Jessome MG, Ronayne ET et al (2022) Understanding the associations among social vulnerabilities, indigenous peoples, and COVID-19 cases within Canadian health regions. Int J Environ Res Public Health 19(19):12409. https://doi.org/10.3390/IJERPH191912409

Jaiswal J, Halkitis PN (2019) Towards a more inclusive and dynamic understanding of medical mistrust informed by science. Behav Med 45(2):79. https://doi.org/10.1080/08964289.2019.1619511

Jayatilleke K (2020) Challenges in implementing surveillance tools of high-income countries (HICs) in low middle income countries (LMICs). Curr Treat Options Infect Dis 12(3):191–201. https://doi.org/10.1007/s40506-020-00229-2

Karbalaei M, Keikha M (2022) Overlapping outbreak of COVID-19 and monkeypox in 2022: warning for immediate preparedness in Iran. Int J Surg 105:106892. https://doi.org/10.1016/J.IJSU.2022.106892

Khankeh H, Farrokhi M, Roudini J, Pourvakhshoori N, Ahmadi S, Abbasabadi-Arab M et al (2021) Challenges to manage pandemic of coronavirus disease (COVID-19) in Iran with a special situation: a qualitative multi-method study. BMC Public Health 21(1):1–9. https://doi.org/10.1186/s12889-021-11973-5

Knight SR, Ho A, Pius R, Buchan I, Carson G, Drake TM et al (2020) Risk stratification of patients admitted to hospital with covid-19 using the ISARIC WHO clinical characterisation protocol: development and validation of the 4C mortality score. BMJ 370:m3339. https://doi.org/10.1136/bmj.m3339

Kretchmer H (2021) Coronavirus: What is vaccine nationalism, how it affects us? World Economic Forum. https://www.weforum.org/agenda/2021/01/what-is-vaccine-nationalism-coronavirus-its-affects-covid-19-pandemic/. Accessed 23 Feb 2023

Kroese K, Porter K, Surridge H, Tembo D (2021) Challenges and solutions: surveying researchers on what type of community engagement and involvement activities are feasible in low and middle income countries during the COVID-19 pandemic. BMJ Open 11(10):e052135. https://doi.org/10.1136/bmjopen-2021-052135

Lee M, You M (2022) Direct and indirect associations of media use with COVID-19 vaccine hesitancy in South Korea: cross-sectional web-based survey. J Med Internet Res 24(1):e32329. https://doi.org/10.2196/32329

Lee VJ, Aguilera X, Heymann D, Wilder-Smith A, Lee VJ, Heymann DL et al (2020) Preparedness for emerging epidemic threats: a lancet infectious diseases commission. Lancet Infect Dis 20(1):17–19. https://doi.org/10.1016/S1473-3099(19)30674-7

Levin-Zamir D, Bertschi I (2018) Media health literacy, eHealth literacy, and the role of the social environment in context. Int J Environ Res Public Health 15(8):1643. https://doi.org/10.3390/ijerph15081643

Lewin S, Munabi-Babigumira S, Glenton C, Daniels K, Bosch-Capblanch X, van Wyk BE et al (2010) Lay health workers in primary and community health care for maternal and child health and the management of infectious diseases. Cochrane Database Syst Rev 2017:CD004015. https://doi.org/10.1002/14651858.CD004015.pub3

Machado FCG, Ferron MM, Barddal MT d M, Nascimento LA, Rosalen J, Avelino-Silva VI (2022) COVID-19 vaccination, incidence, and mortality rates among indigenous populations compared to the general population in Brazil: describing trends over time. Lancet Reg Health Am 13:100319. https://doi.org/10.1016/J.LANA.2022.100319

Matta G (2020) Science communication as a preventative tool in the COVID19 pandemic. Humanit Soc Sci Commun 7(1):159. https://doi.org/10.1057/s41599-020-00645-1

McCool J, Dobson R, Whittaker R, Paton C (2022) Mobile health (mHealth) in low- and middle-income countries. Annu Rev Public Health 43(1):525–539. https://doi.org/10.1146/annurev-publhealth-052620-093850

Mukherjee S (2020) Disparities, desperation, and divisiveness: coping with COVID-19 in India. Psychol Trauma Theory Res Pract Policy 12(6):582. https://doi.org/10.1037/TRA0000682

Muradian R, Arsel M, Pellegrini L, Adaman F, Aguilar B, Agarwal B et al (2013) Payments for ecosystem services and the fatal attraction of win-win solutions. Conserv Lett 6(4):274–279. https://doi.org/10.1111/j.1755-263X.2012.00309.x

Nematollahi S, Minter DJ, Hamaguchi R (2020) Ryoko Hamaguchi, medical student picture of a pandemic: visual aids in the COVID-19 crisis A validated COVID-19 infographic. Resident physician. J Public Health 42(3):483–485. https://doi.org/10.1093/pubmed/fdaa080

Neubeck L, Redfern JU, Fernandez R, Briffa T, Bauman A, Freedman SB (2009) Telehealth interventions for the secondary prevention of coronary heart disease: A systematic review. Eur J Prev Cardiol 16(3):281–289. https://doi.org/10.1097/HJR.0b013e32832a4e7a

Ng R, Tan YW (2022) Media attention toward COVID-19 across 18 countries: the influence of cultural values and pandemic severity. PLoS One 17(12):e0271961. https://doi.org/10.1371/journal.pone.0271961

Nkengasong J, Djoudalbaye B, Maiyegun O (2017) A new public health order for Africa's health security. Lancet Glob Health 5:e1064–e1065. https://doi.org/10.1016/S2214-109X(17)30363-7

Norman CD, Skinner HA (2006) eHealth literacy: essential skills for consumer health in a networked world. J Med Internet Res 8:e9. https://doi.org/10.2196/jmir.8.2.e9

Pallas SW, Minhas D, Pérez-Escamilla R, Taylor L, Curry L, Bradley EH (2013) Community health workers in low- and middle-income countries: what do we know about scaling up and sustainability? Am J Public Health 103(7):e74–e82. https://doi.org/10.2105/AJPH.2012.301102

Paradies Y, Ben J, Denson N, Elias A, Priest N, Pieterse A et al (2015) Racism as a determinant of health: a systematic review and meta-analysis. PLoS One 10(9):e0138511. https://doi.org/10.1371/journal.pone.0138511

Pillay Y, Pienaar S, Barron P, Zondi T (2021) Impact of COVID-19 on routine primary healthcare services in South Africa. S Afr Med J 111(8):714–719. https://doi.org/10.7196/SAMJ.2021.V111I8.15786

Riaz MMA, Ahmad U, Mohan A, dos Santos Costa AC, Khan H, Babar MS et al (2021) Global impact of vaccine nationalism during COVID-19 pandemic. Trop Med Health 49(1):101. https://doi.org/10.1186/s41182-021-00394-0

Rodriguez-Morales AJ, Lopardo G, Verbanaz S, Orduna T, Lloveras S, Azeñas-Burgoa JM et al (2022) Latin America: situation and preparedness facing the multi-country human monkeypox outbreak. Lancet Reg

Health Am 13:100318. https://doi.org/10.1016/j.lana.2022.100318
Rupprecht CE, Burgess GW (2015) Viral and vector zoonotic exploitation of a homo-sociome memetic complex. Clin Microbiol Infect 21(5):394–403. https://doi.org/10.1016/j.cmi.2015.02.032
Sachs JD, Karim SSA, Aknin L, Allen J, Brosbøl K, Colombo F et al (2022) The lancet commission on lessons for the future from the COVID-19 pandemic. Lancet 400(10359):1224–1280. https://doi.org/10.1016/S0140-6736(22)01585-9
Selvarajah S, Maioli SC, Deivanayagam TA, De P, Sato M, Devakumar D et al (2022) Racism, xenophobia, discrimination, and health 2 racism, xenophobia, and discrimination: mapping pathways to health outcomes. Lancet 400:2109–2124. https://doi.org/10.1016/S0140-6736(22)02484-9
Sepúlveda-Loyola W, Rodríguez-Sánchez I, Pérez-Rodríguez P, Ganz F, Torralba R, Oliveira DV, Rodríguez-Mañas L (2020) Impact of social isolation due to COVID-19 on health in older people: mental and physical effects and recommendations. J Nutr Health Aging 24:938–947. https://doi.org/10.1007/s12603-020-1469-2
Shannon G, Morgan R, Zeinali Z, Brady L, Thereza Couto M, Devakumar D et al (2022) Racism, xenophobia, discrimination, and health 3 intersectional insights into racism and health: not just a question of identity. Lancet 400:2125–2136. https://doi.org/10.1016/S0140-6736(22)02304-2
Simpson S, Kaufmann MC, Glozman V, Chakrabarti A (2020) Disease X: accelerating the development of medical countermeasures for the next pandemic. Lancet Infect Dis 20(5):e108–e115. https://doi.org/10.1016/S1473-3099(20)30123-7
Sorvillo TE, Rodriguez SE, Hudson P, Carey M, Rodriguez LL, Spiropoulou CF et al (2020) Towards a sustainable one health approach to Crimean-Congo hemorrhagic fever prevention: focus areas and gaps in knowledge. Trop Med Infect Dis 5(3):113. https://doi.org/10.3390/tropicalmed5030113
Suarez-Lledo V, Alvarez-Galvez J (2021) Prevalence of health misinformation on social media: systematic review. J Med Internet Res 23(1):e17187. https://doi.org/10.2196/17187
Sylvia Chou W-Y, Gaysynsky A, Cappella JN (2020) Where we go from here: health misinformation on social media. Am J Public Health 110(S3):S273–S275. https://doi.org/10.2105/AJPH.2020.305905
Tagoe ET, Sheikh N, Morton A, Nonvignon J, Sarker AR, Williams L, Megiddo I (2021) COVID-19 vaccination in lower-middle income countries: national stakeholder views on challenges, barriers, and potential solutions. Front Public Health 9:709127. https://doi.org/10.3389/fpubh.2021.709127
Tamang ST, Dorji T (2022) Bhutan's preparedness for monkeypox outbreak. Lancet Reg Health Southeast Asia 7:100092. https://doi.org/10.1016/j.lansea.2022.100092
Trolle H, Forsberg B, King C, Akande O, Ayres S, Alfvén T, Elimian K (2023) A scoping review of facilitators and barriers influencing the implementation of surveillance and oral cholera vaccine interventions for cholera control in lower- and middle-income countries. BMC Public Health 23(1):455. https://doi.org/10.1186/S12889-023-15326-2
Tusabe F, Tahir IM, Akpa CI, Mtaki V, Baryamujura J, Kamau B et al (2022) Lessons learned from the Ebola virus disease and COVID-19 preparedness to respond to the human Monkeypox virus outbreak in low- and middle-income countries. Infect Drug Resist 15:6279–6286. https://doi.org/10.2147/IDR.S384348
United Nations. (2021) Goal 3| Department of Economic and Social Affairs [Online].. https://sdgs.un.org/goals/goal3. Accessed 28 Feb 2023
Waitoki W, McLachlan A (2022) Indigenous Māori responses to COVID-19: he waka eke noa? Int J Psychol 57(5):567. https://doi.org/10.1002/IJOP.12849
Wallis L, Blessing P, Dalwai M, Shin SD (2017) Integrating mHealth at point of care in low- and middle-income settings: the system perspective. Glob Health Action 10(sup3):1327686. https://doi.org/10.1080/16549716.2017.1327686
Wang C, Tee M, Roy AE, Fardin MA, Srichokchatchawan W, Habib HA et al (2021) The impact of COVID-19 pandemic on physical and mental health of Asians: A study of seven middle-income countries in Asia. PLoS One 16(2):e0246824. https://doi.org/10.1371/journal.pone.0246824
Wichmann B, Wichmann R (2022) COVID-19 and indigenous health in the Brazilian Amazon. Econ Model 115:105962. https://doi.org/10.1016/j.econmod.2022.105962
World Health Organization (2020) Prioritizing diseases for research and development in emergency contexts. World-Health-Organization. https://www.who.int/activities/prioritizing-diseases-for-research-and-development-in-emergency-contexts
Yinka-Ogunleye A, Aruna O, Dalhat M, Ogoina D, McCollum A, Disu Y et al (2019) Outbreak of human monkeypox in Nigeria in 2017–18: a clinical and epidemiological report. Lancet Infect Dis 19(8):872–879. https://doi.org/10.1016/S1473-3099(19)30294-4

7 Mental Health and SDG3.3, 3d, 3.4 and 3.5

Abstract

Human health is composed of both mental, physical, and social health. Thus, health and wellbeing are measured by the status of these three categories of health. Of the three forms of health, mental health is often neglected and may not receive as much attention as physical and social health. Despite being neglected, mental health plays a major role in both physical and social health. Mental illness has often resulted in discrimination and exclusion of those affected; seriously affecting their social interaction and connectivity. Many suffering from mental disorders tend to withdraw from society further affecting their social health.

Uncertainties brought about by infectious disease outbreaks can trigger distress within a population. Thus management of infectious disease outbreaks should include a mental health component for the affected populations. Pandemics and severe disease outbreaks such as COVID-19, HIV/AIDS, Ebola and other acute but severe disease outbreaks are a trigger for mental health disorders in LMIC. As such culturally appropriate strategies should be set aside to manage mental health issues that come with disease outbreaks of such magnitude.

Keywords

Mental health · Low and Middle Income Countries (LMIC) · COVID-19 · Sustainable development goals · Poverty · Livelihoods · Infectious disease · HIV/AIDS · Ebola

Mental health is a growing public health issue globally. While high income countries (HIC) have invested in management and treatment of various mental health aspects, understanding, treatment and management of mental health are still in their infancy in LMIC. SDG 3.4 focusses on reducing premature mortality from NCD through prevention and treatment and promote mental health and well-being ("By 2030, reduce by one third premature mortality from non-communicable diseases through prevention and treatment and promote mental health and well-being"). Interestingly, mental health is associated with infectious diseases (SDG3.3) as well as substance abuse (SDG 3.5) which have been shown to contribute to mental health outcomes. Building sustainable and resilient mental health systems and setting up infrastructure (SDG3d) to manage these health challenges is essential if LMIC are to achieve SDG3. This chapter will focus on mental health and its associated health outcomes in light with various factors that could positively and or

F. A. Mphande, *Sustainable Health in Low and Middle Income Countries*, Sustainable Development Goals Series, https://doi.org/10.1007/978-981-99-4254-1_7

negatively affect the chances of attaining universal health coverage (SDG 3.8).

7.1 Mental Health and SDG3.3 and 3d

Human health is composed of both mental, physical, and social health. Thus, health and wellbeing are measured by the status of these three categories of health. Of the three forms of health, mental health is often neglected and may not receive as much attention as physical and social health. Despite being neglected, mental health plays a major role in both physical and social health. Mental illness has often resulted in discrimination and exclusion of those affected; seriously affecting their social interaction and connectivity. Many suffering from mental disorders tend to withdraw from society further affecting their social health.

Since the inclusion of mental health in the sustainable development goals, strides have been made in improving mental healthcare globally. Although research funding is still limited, commitments have been made by various organisations to prevent, treat, and research on mental health disorders to improve health and provide better health for all. As of 2019, the highest burden of mental health disorders was reported in Australasia, Tropical Latin America, and high-income North America (GBD 2019 Mental Disorders Collaborators 2022).

Various patterns have emerged across individual disorders as well as regional patterns. The prevalence of depressive disorders was high in sub-Saharan Africa (4540·4 cases per 100,000 people [95% UI 4038·1–5112·4]) and north Africa and the Middle East (4348·9 cases per 100,000 people [3807·3–4971·1]) in addition to Australasia, Tropical Latin America, and high-income North America. The age-standardised prevalence of eating disorders, ADHD, conduct disorder, and autism spectrum disorders was highest in high-income regions. Bipolar disorder and schizophrenia prevalence varied to a lesser extent across regions (GBD 2019 Mental Disorders Collaborators 2022).

SDG3 and 3d state that "Ensuring healthy lives and promoting well-being at all ages is essential to sustainable development" and "Strengthen the capacity of all countries, in particular developing countries, for early warning, risk reduction and management of national and global health risks".

According to WHO, the concept of mental health is described as "state of well-being in which the individual realizes his or her own abilities, can cope with the normal stresses of life, can work productively and fruitfully, and is able to make a contribution to his or her community" (https://www.paho.org/en/topics/mental-health). Mental health is crucial in achieving SDG3 and 3d globally and specifically in LMIC. Mental health affects people of all ages and is the leading cause of depression, and severe mental health can lead to death (https://www.who.int/health-topics/mental-health#tab=tab_1).Despite mental health and associated conditions being accepted in some parts of the world, discrimination, abuse and stigma are observed especially in countries where the condition is not widely accepted as a health issue (Banerjee et al. 2021; Henderson et al. 2014; Mukherjee 2020). In LMIC support for as well as treatment of mental health issues are still in their infancy. While some LMIC have policies on mental health, others are still lagging, leading to poor management and treatment of mental health patients. As part of SDG, universal health coverage is a goal that countries are striving to achieve globally. As such integration of mental healthcare into existing health systems in LMIC and globally is essential.

In the past 2 years, mental healthcare and management was further complicated by the COVID-19 pandemic affecting both the treatment as well as the progress that was made in some of the mental health patients (Khan et al. 2020; McClarty et al. 2022; Sepúlveda-Loyola et al. 2020).

Mental health is among the top 10 leading causes of disease burden worldwide with the number of DALYs increasing from 80.8 million to 125.3 million between 1990 and 2019 respectively (GBD 2019 Mental Disorders Collaborators 2022). This increase has been associated with economic losses both in high income and low-income countries with approximately US $5 trillion associated with mental health burden. The losses account for up to 4% and 8% gross domestic product (GDP) in Eastern Sub-Saharan Africa and High-income North America (Arias et al. 2022). Mental health disorders are grossly underestimated even though these disorders cause significant morbidity and mortality in affected populations globally. Mental health disorders can lead to disability, and contribute to poverty and premature death in affected populations (Vigo et al. 2016). Mental health disorders have been associated with substance use and vulnerable populations and LMIC carry most of the burden (Arias et al. 2022; Charlson et al. 2015; Vigo et al. 2016). Despite the greater burden, mental health is often neglected in many countries and some of the barriers include, limited funding, shortage of skills and skilled personnel to manage the burden, stigma associated with mental illness and limited financing for mental health research (Hall et al. 2019; Mukherjee 2020; Tora et al. 2018).

Mental health has been linked to poverty due to the economic consequences as the result of disability, loss of livelihood, absenteeism, costs for medical care (hospital visits, medications, therapy. Trautmann et al. reported that costs for mental disorders were had surpassed those of cancer and diabetes (Trautmann et al. 2016).

Lower- and middle-income countries are burdened by infectious diseases, chronic non-communicable diseases, poverty, vulnerable populations, health emergencies and natural disasters, as well as struggling healthcare systems both of which could contribute to mental health issues (Fig. 7.1). It is crucial to consider SDG3.4 "By 2030, reduce by one third premature mortality from non-communicable diseases through prevention and treatment and promote mental health and well-being" when managing mental health. Non communicable diseases (NCD) have contributed to mental health in individuals due to their chronic nature and the negative effects of the disease. Similarly, mental health is an NCD and thus integration of mental health services in routine healthcare could assist in prevention and treatment of mental health disorders (Trivedi 2017).

7.2 Mental Health in LMIC

In 2021, the population of LMIC was estimated at approximately 3.4 billion which represents 43% of the world population (The World Bank n.d.). Lower- and middle-income countries present varying settings of healthcare, social political environment, and geographical environment. These play a crucial role in the health of the populations in LMIC. Governance, emergencies, and unstable political environments have created situations that have left populations in these regions vulnerable to various mental health challenges including post traumatic disorder (PTSD), depression, anxiety, and many other mental health challenges. Over 70% of mental illness occur in LMIC with humanitarian crises as one of the major contributors of mental health issues in the region (Bürgin et al. 1974, 2022; GBD 2019 Mental Disorders Collaborators 2022; Punamäki et al. 2015; Rathod et al. 2017). Humanitarian crises is a broad term used to refer to rapidly deteriorating situation Humanitarian crises are commonly defined as rapid and serious deteriorations in safety, with numerous victims threatening lives, and security of a large number of people or numerous people associated with distress, and displacement. The most commonly affected populations live in low- and middle-income countries (Kamadjeu et al. 2014; Kelly-Hope et al. n.d.; Kevlihan 2013; Purgato et al. 2018; Zakham et al. 2019). Humanitarian crises can be in form of natural disasters, infectious diseases and or any other form of emergency that results in loss of lives, property, livelihood, and unsanitary living conditions. These conditions are a breeding ground for infectious disease outbreak that further complicate the already strenuous situation. The strain brought about by humanitarian crises affects the mental health of affected populations both young and old (Abbas et al. 2017; Greene et al. 2017; Kamadjeu et al. 2014; Kelly-Hope et al. n.d.). Apart from humanitarian crises affecting mental health, crises affect health coverage.

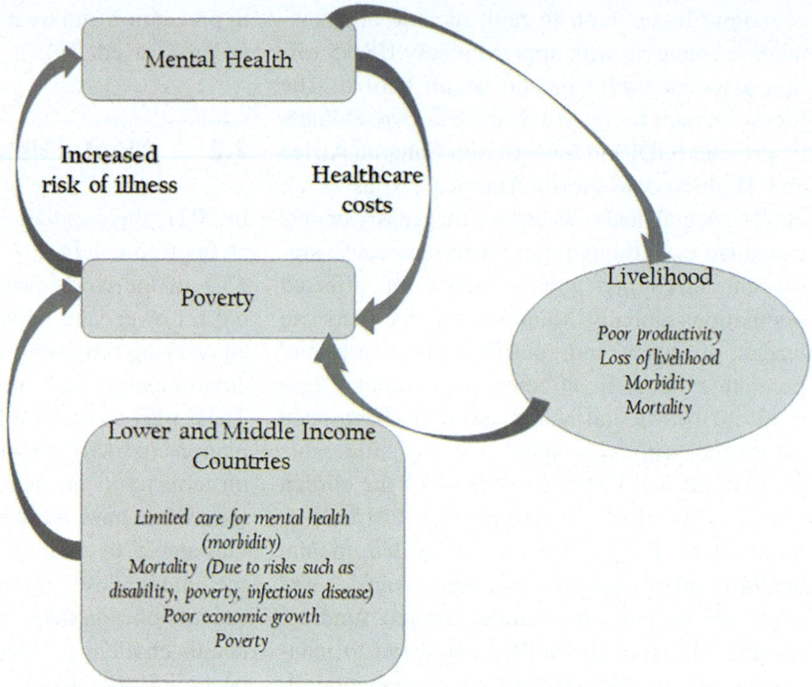

Fig. 7.1 The interconnection between poverty, livelihoods in low- and middle-income countries and how these contribute to mental health

In settings affected by war, disaster and other forms of emergencies, health services are often disrupted negatively affecting universal health coverage both in LMIC and globally (SDG3.8 Achieve universal health coverage, including financial risk protection, access to quality essential health-care services and access to safe, effective, quality and affordable essential medicines and vaccines for all). Thus there is need to ensure that populations in humanitarian crises have access to healthcare including medicines, vaccines and mental health services (Marie Knaul et al. 2018).

Literature has shown that there is a link between poverty and mental health. Stressors that come with economic hardships including poor living conditions, limited and or lack of income play a role in associated health outcomes. There link between mental health, livelihood, poverty, and diseases is illustrated (Fig. 7.2).

Mental health situations are perceived differently in different parts of the world. These perceptions are moulded by culture, setting and social values (Kohrt and Hruschka 2010; Wessells 2009). Different countries view mental health through the lens of their norms and values. While in some countries mental health can be discussed openly, it is a taboo topic in others hence interventions have to be carried out with proper consultation of communities involved (Checchi et al. 2017; Ganesan 2006; Greene et al. 2017; Kohrt and Hruschka 2010; Purgato et al. 2018; Wessells 2009). These perceptions could affect how mental health is viewed and managed in various contexts. Mental health has been associated with several factors including intimate partner violence such as physical, sexual and psychological abuse and substance abuse (García-Moreno et al. 2015; Tol et al. 2019). For most LMIC, noncommunicable diseases, infectious diseases, and poverty have contributed to mental health disorders. Poor health from disease directly can affect an individual's mental health as they try to manage the healthcare costs brought about by these diseases as well as costs to manage the mental health disorders that could be a result of mental health treatment (Lund 2014; Sabes-Figuera et al. 2012; Vigo et al. 2016) (Fig. 7.2). The management of infectious and NCDs in LMIC is already enormous, although in its infancy in comparison to high income countries, mental healthcare is a crucial economic need in LMIC

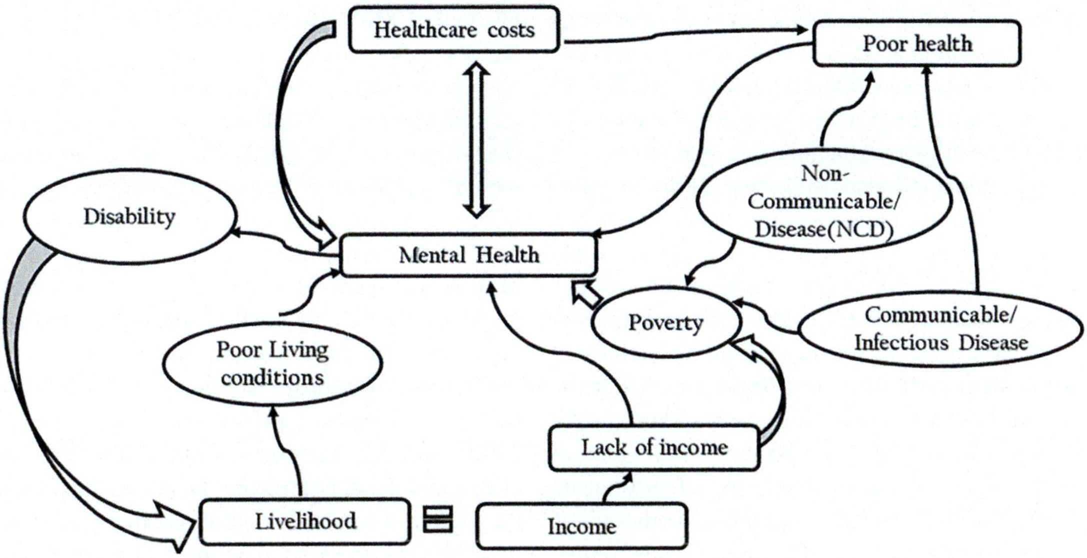

Fig. 7.2 Factors that contribute to mental health, stressors are indicated with thick arrows, these include healthcare costs, disability, poverty, lack of income and mental health; factors that contribute to the stressors are shown in line arrows

that should not be neglected if SDG3.4 is to be achieved in this region.

A study by Koschorke et al., on seven countries in Africa, Asia, and Europe showed that there are several players in management of mental healthcare, these include primary care providers, individuals with mental illness, family and community members (Koschorke et al. 2021). While some countries have well trained primary care specialists for mental health, most countries particularly LMIC have limited at times no specialist training (Koschorke et al. 2021). As such the way healthcare workers view and manage mental health patients is often different (Koschorke et al. 2021).

Some of the barriers to mental health include stereotypes and negative beliefs, intentional and unintentional discrimination, and stigma (Drew et al. 2011; Henderson et al. 2014; Ross and Goldner 2009; Vistorte et al. 2018). Individuals going through stressful and or distressing situations are prone to suffer mental health illness. In trying to cope, some may be caught up in substance abuse including drugs and alcohol, while others may commit suicide (Banerjee et al. 2021; Global Burden of Disease Study 2013 Collaborators 2015; Vigo et al. 2016). Substance abuse has been associated with stigma in LMIC making it difficult for affected individuals to get treatment (Banerjee et al. 2021; Rathod et al. 2017).

7.2.1 Risks and Determinants of Mental Health in LMIC

Psychiatric and physical disorders contribute to mental health and wellbeing globally. Psychiatric disorders including schizophrenia, bipolar affective disorder, and recurrent major depressive disorder have been associated with mental health both in the young and old. Determinants of mental health include individuals' genetic disposition, psychological and biological factors, and substance use. Exposure to stressful as well as distressing circumstances including economic challenges, wars and displacement, poverty, health status, inequalities and oppression, and many other social, political and economic issues that could trigger distress and anxiety (Bürgin et al. 2022; Khan et al. 2020; Li and Loke 2013; Mousley et al. 2014; Sabes-Figuera et al. 2012;

Tora et al. 2018). Risk factors may be as early as from childhood abuse and bullying to social economic factors in adulthood (https://www.who.int/news-room/fact-sheets/detail/mental-health-strengthening-our-response).

Individual attributes including ability to maintain positive social interactions, environment and economic well-being are protective factors, while displacement, forced migration, political upheavals and sudden loss (disasters and emergencies) negatively affect mental health outcomes (https://www.who.int/news-room/fact-sheets/detail/mental-health-strengthening-our-response). As individuals struggle to manage their mental health, some turn to excessive use of substances such as alcohol, cannabis, opioids and other illicit drugs which negatively affect mental health outcomes. Harmful substances such as alcohol, opioids, and cannabis have been associated with mental disorders (Hunt et al. 2016; Koskinen et al. 2010).

Physical disorders such as NCD and infectious diseases have also been linked to mental disorders. Mortality among people with severe mental illness has been associated with comorbidities including NCD obesity and hypertension cardiovascular disease, respiratory diseases, cancers and infectious diseases including syphilis and HIV as well as substance abuse (Di Florio et al. n.d.; Laursen et al. 2012; Lawrence et al. 2013; Mpango et al. 2023; Smith et al. 2013).

Management of mental health in LMIC varies in different settings. Collaborative partnerships between HIC and LMIC have been ongoing although inequitable knowledge sharing has been observed. To tackle this, collaborations between LMIC through the Programme for Improving Mental Health Care (PRIME), initiative was launched in several LMIC. The partnership led by LMIC, aimed to provide research evidence for the development, implementation and scaling up of integrated district mental healthcare plans in Ethiopia, India, Nepal, South Africa and Uganda (Breuer et al. 2019). Cultural values, norms and beliefs play a crucial role in how mental health issues are addressed in different societies, as such great consideration should be taken when creating policies with research evidence to improve health in LMIC (Breuer et al. 2019). Systemic, factors, lack of awareness, as well as inequalities have further affected management of severe mental disorders by healthcare providers in LMIC (Das-Munshi et al. 2020). As such to improve mental healthcare outcomes in LMIC awareness in both communities and healthcare providers is required to ensure that there is ample awareness of mental health issues to ensure appropriate response. By training health personnel and building systems that integrate mental health in routine primary healthcare would assist in achieving better outcomes in mental health in LMIC. Behavioural intervention studies are crucial in understanding mental health challenges at the grassroot to identify solutions that can assist in achieving positive outcomes (Bryant et al. 2022). Developing customized tools for mental health education and awareness, well suited for respective communities is crucial in attaining positive outcomes (Duara et al. 2022).

7.2.2 Infectious Disease and Mental Health

Various infectious disease outbreaks have been associated with mental health due to the related health outcomes. Diseases such as Ebola virus disease (EVD), Zika, SARS, H1N1, and COVID-19 have resulted in post-traumatic disorders (PTSD) in some survivors and those affected by the disease (Banerjee et al. 2021; Jacobs-Wingo et al. 2016; Punamäki et al. 2015; Slone et al. 2017; Sprecher et al. 2017; Stranix-Chibanda et al. 2005) (Fig. 7.3). Infectious disease outbreaks can elicit varying emotional reactions in vulnerable groups and populations. Brooks et al. reported negative emotional state, living with uncertainty, concerns about the infection disrupted routines, disrupted expectations as well as social support as some of the factors that affected pregnant women during pregnancy (Brooks et al. 2020). Financial and occupational concerns are another set of factors of concern that may affect mental health during infectious disease outbreaks. Individuals may end up quitting high risk jobs, avoid certain workplaces that

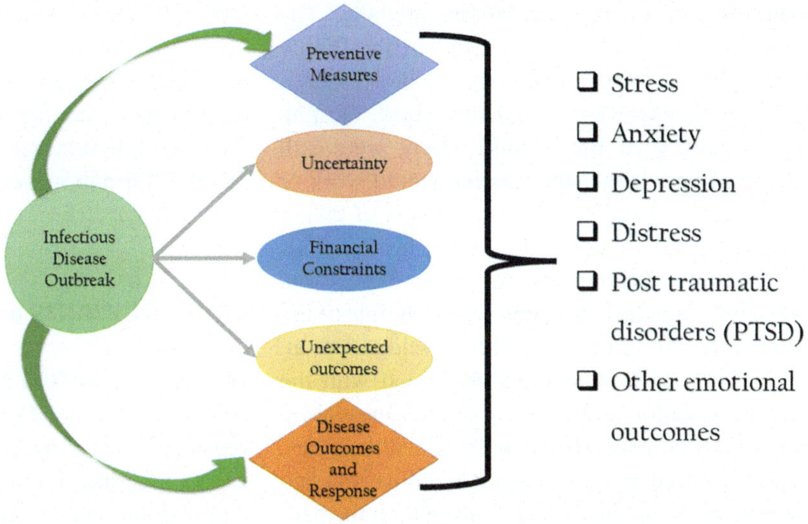

Fig. 7.3 Infectious disease and mental health, infectious disease outbreaks come with unexpected outcomes, uncertainties, financial constraints, as well as preventive measure and disease outcomes that bring about mental health challenges such as stress, anxiety, depression, distress as well as other emotional outcomes

increase the risk of exposure to disease as well as increased expenses for protective equipment such as masks and other safety measures (Lee et al. 2006; Lohm et al. 2014; Ng et al. 2004).

7.2.3 Ebola Virus Disease and Mental Health

Ebola virus disease is a zoonotic disease transmitted from animals to humans with sustained human to human transmission (https://www.who.int/news-room/fact-sheets/detail/ebola-virus-disease?gclid=CjwKCAiAmJGgBhAZEiwA1JZolkAaBTDpVTzBJfRSCVsdjZ9c4h2gGP3wCqf6TNdq_CL84jyrDSJFEBoCeasQAvD_BwE).

With a case fatality rate ranging from 25% to 90% and ana average of 50%, the disease has been associated with fear and distress among the affected populations. Due to the highly infectious nature of the disease and transmissibility, patients suffering from Ebola are separated from their families and often left with healthcare workers; while others are left to die as their families watch and wait for the healthcare workers to come and collect the bodies of their loved ones to bury them. In a culture where family members take care of their sick and bury their loved ones and the associated death, EBV has resulted in psychological distress for those who have experienced the disease (Betancourt et al. 2016; Cénat et al. 2019; Kamara et al. 2017). Anxiety, fear of death, loss of loved ones due to disease and stigma are some of the risk factors that have been identified in survivors, healthcare workers and families (Betancourt et al. 2016; Cénat et al. 2019; Kamara et al. 2017). In Sierra Leone, some of the psychological stressors encountered included inability to acquire property, family rejection, community rejection, and job loss for survivors and associated family members (Bakare et al. 2016). Similarly, in treatment centres in Northern Kivu, it was observed that death of an individual in the treatment camp triggered anxiety in patients receiving treatment and their caregivers (Cénat et al. 2019). Adherence to prevention strategies was observed to trigger anxiety while knowledge of someone suffering from disease was associated to adherence to prevention measure (Betancourt et al. 2016).

7.2.4 COVID-19 and Mental Health

The COVID-19 pandemic presented a high degree of uncertainty, with no clear information on how long the pandemic would last, the high mortality, severity of the disease and whether oneself or their family members would catch the disease and whether they will be able to survive the disease (Golechha 2020; Lai et al. 2021; Tee et al. 2021). This scenario brought a lot of anxiety and distress among populations globally. The high mortality rate that occurred in the highly affected countries in Europe and Americas brought shockwaves around the world. Despite the discovery of the vaccine many were still skeptical on whether the vaccine would work or not and what would be the side effects of taking the vaccine. An increase in mental health needs was reported with many reporting depression, anxiety, suicidal thoughts, and helplessness (Golechha 2020; Lai et al. 2021; Veldhuis et al. 2021). For those who were diagnosed with the disease isolation and duration of the illness were stressors that exacerbated mental health issues. There is need to prioritize availability of, and access to, mental health care during both the pandemic and the recovery (Golechha 2020; Khan et al. 2020; Veldhuis et al. 2021).

COVID-19 had an impact on mental health not only individuals suffering from the disease but also healthcare workers (H. Hall 2020; WHO Health Workforce Department 2021). Growing anxiety was reported in students in LMIC who had challenges to join online courses and or their schools could not provide online education (Afzal et al. 2022; Chowdhury et al. 2022). In university students, the stressors included prospects of graduating, unemployment, as well as financial instability; as families struggled financially due to closure of businesses, loss of jobs and other financial constraints, this weighed heavily on the students some of whom had to take up financial responsibilities for their families (Khan et al. 2020). Studies in several Asian countries using Impact of Event Scale-Revised (IES-R) and Depression, Anxiety and Stress Scale (DASS-21) revealed different levels of depression, anxiety and stress in different countries (Tee et al. 2021; Wang et al. 2021). Significant differences were observed in the scores with Thailand scoring highest on depression, anxiety and stress and Vietnam scoring the lowest (Wang et al. 2021). Risk and protective factors were identified in seven middle income countries (MIC), China, Iran, Malaysia, Pakistan, Philippines, Thailand, and Vietnam. It was observed that age (below 30 years), education background, marital status, discrimination and contact with individuals with COVID-19 were risk factors for adverse mental health. While the male gender, living with children and other individuals in the same household, employment, confidence in doctors, perceived likelihood of survival and spending less time on health information were protective factors (Wang et al. 2021).

7.2.5 HIV/AIDS and Mental Health

UNAIDS report 2022 has revealed changes in the progress towards HIV/AIDS including shrinking resources, widening inequalities, and 1.5 million new infections, 1 million more than the global target (Joint United Nations Programme on HIV/AIDS 2022). Gaps in HIV treatment access have been observed between high income and low income countries as well as within the countries where certain parts of a country have better access compared to others (Joint United Nations Programme on HIV/AIDS 2022). Differences have also been observed in access to treatment between children and adults, with adults having more access than children. It has been reported that over 1.5 million children are born to HIV seropositive (HIV+) mothers each year thanks to the prevention of mother to child transmission of the virus (Connor et al. 1994; Kapetanovic et al. 2014). However, despite these efforts, HIV+ women have been affected with depression and other mental health related disorders. Depression, anxiety and distress were reported in LMIC; while some women had episodes of loss of interest in life, ideas of worthlessness others experienced severe anxiety and distress (Kapetanovic et al. 2014; Stranix-Chibanda et al. 2005). Some of the worry from HIV+ mothers was for their HIV+ babies as well as inability to disclose to others their HIV status to avoid stigma (Bennetts et al. 1999).

In adolescents, the COVID-19 pandemic disruptions led to approximately 1 million girls out of school and spikes in teenage pregnancies and gender based violence (Joint United Nations Programme on HIV/AIDS 2022). Suicidal behaviour was reported in some of the pregnant adolescent girls (Musyimi et al. 2020). Depression was also observed as a comorbidity in children affected with HIV(Binagwaho et al. 2021).

7.2.6 Mental Health and SDG3.8 (Universal Health Coverage)

Sustainable development goal 3.8, focusses on universal health coverage for all ("Achieve universal health coverage, including financial risk protection, access to quality essential health-care services and access to safe, effective, quality and affordable essential medicines and vaccines for all"). Mental health in LMIC requires greater attention. While policies have been put in place to manage mental health in LMIC, implementation and sustainability of such policies is essential in attaining universal health coverage (Abbas et al. 2017). Risks to mental health span from social, economic and environmental factors, with both NCD and communicable diseases negatively affecting mental health outcomes (Borges do Nascimento et al. 2022; Bürgin et al. 1974; Jones et al. 2021; Khankeh et al. 2021; Sprecher et al. 2017). While programmes are in place to manage various NCD and communicable diseases it is crucial that such initiatives should be considered for mental health. Health systems in LMIC are burdened with NCD and communicable diseases which have overshadowed the burden of mental health issues. Developing infrastructure that could accommodate and address NCD, communicable diseases and mental health issues on the same level could assist achieving equitable health and subsequently universal health coverage (Bürgin et al. 2022; Hall et al. 2019; Lawrence et al. 2013; McClarty et al. 2022; Rathod et al. 2017; Sabes-Figuera et al. 2012; Stranix-Chibanda et al. 2005). Costs involved in mental health management in the long term are a challenge for populations in LMIC. Stressors such as health emergencies, environmental disasters, wars, and political instability have raised the need for psychological support in the affected communities. Lack of finances, trained personnel, infrastructure and skills support have rendered mental health services almost non-existent in most communities and health systems in LMIC (Bürgin et al. 2022). Attaining SDG3d (establishment of resilient infrastructure), would improve chances of attaining SDG3.4, SDG3.5 and SDG3.8 all of which require sustainable infrastructure to provide services to the populations. Thus, achieving universal health coverage means achieving targets and indicators from SDG3.1 to SDG3.7, 3a, 3b, 3c and 3d after which SDG3.9 could be attainable (https://sdgs.un.org/goals/goal3).

7.3 Conclusion

Mental health affects people of all ages and is the leading cause of depression, and in case of severe mental health death. The burden of infectious disease outbreaks in LMIC and their role in mental health should not be ignored. If SDG3 all its targets are to be achieved in LMIC, mental health must be given a priority similar to that given to communicable diseases and other NCD. The vicious cycle of mental health, livelihood, and poverty in LMIC should be carefully considered when designing infectious disease policies in the region. The challenges faced by LMIC to meet healthcare costs amidst struggling healthcare systems, the burden of NCD, infectious disease, and mental health present a huge hurdle in achieving SDG3 and its components, including universal health coverage. Finding means to strengthen capacities in LMIC and combating poverty could reduce the burden of infectious diseases and mental health. Uncertainties brought about by infectious disease outbreaks can trigger distress within a population. Thus, management of infectious disease outbreaks and other health emergencies should include a mental health component for the affected populations. Pandemics and severe disease outbreaks such as CoVID-19, HIV/AIDS, Ebola and other acute but severe disease outbreaks are a trigger for mental health disorders. As such culturally appropriate strategies should be set aside to manage mental health issues associated with disease outbreaks of such magnitude.

References

Abbas M, Aloudat T, Bartolomei J, Carballo M, Durieux-Paillard S, Gabus L et al (2017) Migrant and refugee populations: a public health and policy perspective on a continuing global crisis. Antimicrob Resist Infect Control 7:113. https://doi.org/10.1186/s13756-018-0403-4

Afzal SS, Qamar MA, Dhillon R, Bhura M, Khan MH, Suriya Q et al (2022) Budding medical professionals and COVID-19: the impact of COVID-19 on mental health and medical students. J Ayub Med Coll Abbottabad 34(3):483–488. https://doi.org/10.55519/JAMC-03-9572

Arias D, Saxena S, Verguet S (2022) Quantifying the global burden of mental disorders and their economic value. EClinicalMedicine 54:101675. https://doi.org/10.1016/j.eclinm.2022.101675

Bakare WA, Ilesanmi OS, Nabena EP, Famuyide T (2016) Psychosocial stressors and support needs of survivors of Ebola virus disease, Bombali District, Sierra Leone, 2015. Healthc Low-Resour Settings 3(2):48–51. https://doi.org/10.4081/hls.2015.5411

Banerjee D, Kosagisharaf JR, Sathyanarayana Rao TS (2021) 'The dual pandemic' of suicide and COVID-19: a biopsychosocial narrative of risks and prevention. Psychiatry Res 295:113577. https://doi.org/10.1016/J.PSYCHRES.2020.113577

Bennetts A, Shaffer N, Manopaiboon C, Chaiyakul P, Siriwasin W, Mock P et al (1999) Determinants of depression and HIV-related worry among HIV-positive women who have recently given birth, Bangkok, Thailand. Soc Sci Med 49(6):737–749. https://doi.org/10.1016/S0277-9536(99)00108-2

Betancourt TS, Brennan RT, Vinck P, VanderWeele TJ, Spencer-Walters D, Jeong J et al (2016) Associations between mental health and Ebola-related health behaviors: a regionally representative cross-sectional survey in post-conflict Sierra Leone. PLoS Med 13(8):e1002073. https://doi.org/10.1371/journal.pmed.1002073

Binagwaho A, Remera E, Bayingana AU, Gishoma D, Scott KW, Goosman M et al (2021) Addressing the mental health needs of children affected by HIV in Rwanda: validation of a rapid depression screening tool for children 7–14 years old. BMC Pediatr 21(1):59. https://doi.org/10.1186/s12887-020-02475-1

Borges do Nascimento IJ, Pizarro AB, Almeida JM, Azzopardi-Muscat N, Gonçalves MA, Björklund M, Novillo-Ortiz D (2022) Infodemics and health misinformation: a systematic review of reviews. Bull World Health Organ 100(9):544–561. https://doi.org/10.2471/BLT.21.287654

Breuer E, Hanlon C, Bhana A, Chisholm D, De Silva M, Fekadu A et al (2019) Partnerships in a global mental health research programme—the example of PRIME. Glob Soc Welf 6(3):159. https://doi.org/10.1007/S40609-018-0128-6

Brooks SK, Weston D, Greenberg N (2020) Psychological impact of infectious disease outbreaks on pregnant women: rapid evidence review. Public Health 189:26–36. https://doi.org/10.1016/j.puhe.2020.09.006

Bryant RA, Malik A, Aqel IS, Ghatasheh M, Habashneh R, Dawson KS et al (2022) Effectiveness of a brief group behavioural intervention on psychological distress in young adolescent Syrian refugees: a randomised controlled trial. PLoS Med 19(8):e1004046. https://doi.org/10.1371/JOURNAL.PMED.1004046

Bürgin D, Anagnostopoulos D, Vitiello B, Sukale T, Schmid M, Fegert JM (1974) Impact of war and forced displacement on children's mental health-multilevel, needs-oriented, and trauma-informed approaches. Eur Child Adolesc Psychiatry 31:845–853. https://doi.org/10.1007/s00787-022-01974-z

Bürgin D, Anagnostopoulos D, Doyle M, Eliez S, Fegert J, Fuentes J et al (2022) Impact of war and forced displacement on children's mental health—multilevel, needs-oriented, and trauma-informed approaches. Eur Child Adolesc Psychiatry 31(6):845–853. https://doi.org/10.1007/S00787-022-01974-Z

Cénat JM, Balayulu-Makila O, Derivois D, Felix N, Mukunzi JN, Bukaka J et al (2019) The mental health professional at the feet of the Ebola virus: individual treatment and global mental challenges. Br J Psychiatry 215:577–578. https://doi.org/10.1192/bjp.2019.122

Charlson FJ, Baxter AJ, Dua T, Degenhardt L, Whiteford HA, Vos T (2015) Excess mortality from mental, neurological and substance use disorders in the global burden of disease study 2010. Epidemiol Psychiatr Sci 24(2):121–140. https://doi.org/10.1017/S2045796014000687

Checchi F, Warsame A, Treacy-Wong V, Polonsky J, van Ommeren M, Prudhon C (2017) Public health information in crisis-affected populations: a review of methods and their use for advocacy and action. Lancet 390(10109):2297–2313. https://doi.org/10.1016/S0140-6736(17)30702-X

Chowdhury U, Suvro MAH, Farhan SMD, Uddin MJ (2022) Depression and stress regarding future career among university students during COVID-19 pandemic. PLoS One 17(4):e0266686. https://doi.org/10.1371/journal.pone.0266686

Connor EM, Sperling RS, Gelber R, Kiselev P, Scott G, O'Sullivan MJ et al (1994) Reduction of maternal-infant transmission of human immunodeficiency virus type 1 with zidovudine treatment. Pediatric AIDS clinical trials group protocol 076 study group. N Engl J Med 331(18):1173–1180. https://doi.org/10.1056/NEJM199411033311801

Das-Munshi J, Semrau M, Barbui C, Chowdhary N, Gronholm PC, Kolappa K et al (2020) Gaps and challenges: WHO treatment recommendations for tobacco cessation and management of substance use disorders in people with severe mental illness. BMC Psychiatry 20(1):237. https://doi.org/10.1186/s12888-020-02623-y

Di Florio A, Craddock N, Van Den Bree M (n.d.) Alcohol misuse in bipolar disorder. A systematic review and meta-analysis of comorbidity rates. Eur Psychiatry 29(3):117–124. https://doi.org/10.1016/j.eurpsy.2013.07.004

Drew N, Funk M, Tang S, Lamichhane J, Chávez E, Katontoka S et al (2011) Human rights violations of people with mental and psychosocial disabilities: an unresolved global crisis. Lancet 378(9803):1664–1675. https://doi.org/10.1016/S0140-6736(11)61458-X

Duara R, Chowdhury D, Dey R, Goswami S, Madill A (2022) Using cocreated visually informed community mental health education in low- and middle-income countries: a case study of youth substance misuse in Assam, India. Health Expect 25(4):1930–1944. https://doi.org/10.1111/hex.13550

Ganesan M (2006) Psychosocial response to disasters—some concerns. Int Rev Psychiatry 18(3):241–247. https://doi.org/10.1080/09540260600656035

García-Moreno C, Zimmerman C, Morris-Gehring A, Heise L, Amin A, Abrahams N et al (2015) Addressing violence against women: a call to action. Lancet 385(9978):1685–1695. https://doi.org/10.1016/S0140-6736(14)61830-4

GBD 2019 Mental Disorders Collaborators (2022) Global, regional, and national burden of 12 mental disorders in 204 countries and territories, 1990–2019: a systematic analysis for the global burden of disease study 2019. Lancet Psychiatry 9(2):137–150. https://doi.org/10.1016/S2215-0366(21)00395-3

Global Burden of Disease Study 2013 Collaborators (2015) Global, regional, and national incidence, prevalence, and years lived with disability for 301 acute and chronic diseases and injuries in 188 countries, 1990-2013: a systematic analysis for the global burden of disease study 2013. Lancet (London, England) 6736(15):1990–2013. https://doi.org/10.1016/S0140-6736(15)60692-4

Golechha M (2020) COVID-19, India, lockdown and psychosocial challenges: what next? Int J Soc Psychiatry 66(8):830. https://doi.org/10.1177/0020764020935922

Greene MC, Jordans MJD, Kohrt BA, Ventevogel P, Kirmayer LJ, Hassan G et al (2017) Addressing culture and context in humanitarian response: preparing desk reviews to inform mental health and psychosocial support. Confl Heal 11(1):21. https://doi.org/10.1186/s13031-017-0123-z

Hall H (2020) The effect of the COVID-19 pandemic on healthcare workers' mental health. JAAPA 33(7):45–48. https://doi.org/10.1097/01.JAA.0000669772.78848.8c

Hall T, Kakuma R, Palmer L, Martins J, Minas H, Kermode M (2019) Are people-centred mental health services acceptable and feasible in Timor-Leste? A qualitative study. Health Policy Plan 34(Suppl. 2):II93–II103. https://doi.org/10.1093/HEAPOL/CZZ108

Henderson C, Noblett J, Parke H, Clement S, Caffrey A, Gale-Grant O et al (2014) Mental health-related stigma in health care and mental health-care settings. Lancet Psychiatry 1(6):467–482. https://doi.org/10.1016/S2215-0366(14)00023-6

Hunt GE, Malhi GS, Cleary M, Man H, Lai X, Sitharthan T (2016) Comorbidity of bipolar and substance use disorders in national surveys of general populations, 1990–2015: systematic review and meta-analysis. J Affect Disord 206:321–330. https://doi.org/10.1016/j.jad.2016.06.051

Jacobs-Wingo J, Ezeoke I, Saffa A, Tate A, Lee D, Johnson K et al (2016) Using a call center to coordinate Zika virus testing-New York city, 2016. J Emer Manag 14(6):391–395. https://doi.org/10.5055/jem.2016.0303

Joint United Nations Programme on HIV/AIDS. (2022) UNAIDS Global AIDS Update 2022. *UNAIDS Global AIDS Update 2022, 2022*, 1–22

Jones EAK, Mitra AK, Bhuiyan AR (2021) Impact of COVID-19 on mental health in adolescents: a systematic review. Int J Environ Res Public Health 18(5):2470. https://doi.org/10.3390/ijerph18052470

Kamadjeu R, Mahamud A, Webeck J, Baranyikwa MT, Chatterjee A, Bile YN et al (2014) Polio outbreak investigation and response in Somalia, 2013. J Infect Dis 210(Suppl):S181–S186. https://doi.org/10.1093/infdis/jiu453

Kamara S, Walder A, Duncan J, Kabbedijk A, Hughes P, Muana A (2017) Mental health care during the Ebola virus disease outbreak in Sierra Leone. Bull World Health Organ 95(12):842–847. https://doi.org/10.2471/BLT.16.190470

Kapetanovic S, Dass-Brailsford P, Nora D, Talisman N (2014) Mental health of HIV-seropositive women during pregnancy and postpartum period: a comprehensive literature review. AIDS Behav 18(6):1152–1173. https://doi.org/10.1007/s10461-014-0728-9

Kelly-Hope LA, Sanders AM, Harding-Esch E, Willems J, Ahmed F, Vincer F, Hill R (n.d.) Complex emergencies and the control and elimination of neglected tropical diseases in Africa: developing a practical approach for implementing safe and effective mapping and intervention strategies. Confl Heal 15(1):18. https://doi.org/10.1186/s13031-021-00356-7

Kevlihan R (2013) Providing health services during a civil war: the experience of a garrison town in South Sudan. Disasters 37(4):579–603. https://doi.org/10.1111/disa.12027

Khan AH, Sultana MS, Hossain S, Hasan MT, Ahmed HU, Sikder MT (2020) The impact of COVID-19 pandemic on mental health & wellbeing among home-quarantined Bangladeshi students: a cross-sectional pilot study. J Affect Disord 277:121–128. https://doi.org/10.1016/j.jad.2020.07.135

Khankeh H, Farrokhi M, Roudini J, Pourvakhshoori N, Ahmadi S, Abbasabadi-Arab M et al (2021) Challenges to manage pandemic of coronavirus disease (COVID-19) in Iran with a special situation: a qualitative multi-method study. BMC Public Health 21(1):1–9. https://doi.org/10.1186/s12889-021-11973-5

Kohrt BA, Hruschka DJ (2010) Nepali concepts of psychological trauma: the role of idioms of distress, ethnopsychology and ethnophysiology in alleviating suffering and preventing stigma. Cult Med Psychiatry 34(2):322–352. https://doi.org/10.1007/s11013-010-9170-2

Koschorke M, Oexle N, Ouali U, Cherian AV, Deepika V, Mendon GB et al (2021) Perspectives of healthcare providers, service users, and family members about mental illness stigma in primary care settings: a multi-site qualitative study of seven countries in Africa, Asia, and Europe. PLoS One 16(10):e0258729. https://doi.org/10.1371/journal.pone.0258729

Koskinen J, Löhönen J, Koponen H, Isohanni M, Miettunen J (2010) Rate of cannabis use disorders in clinical samples of patients with schizophrenia: a meta-analysis. Schizophr Bull 36(6):1115. https://doi.org/10.1093/SCHBUL/SBP031

Lai J, Jain T, Sangraula M, Tee M, Wang C, Tee C et al (2021) Impact of the COVID-19 pandemic on physical and mental health in lower and upper middle-income Asian countries: a comparison between the Philippines and China. Front Psych 11:568929. https://doi.org/10.3389/fpsyt.2020.568929

Laursen TM, Munk-Olsen T, Vestergaard M (2012) Life expectancy and cardiovascular mortality in persons with schizophrenia. Curr Opin Psychiatry 25(2):83–88. https://doi.org/10.1097/YCO.0B013E32835035CA

Lawrence D, Hancock KJ, Kisely S (2013) The gap in life expectancy from preventable physical illness in psychiatric patients in Western Australia: retrospective analysis of population based registers. BMJ 346(7909):f2539. https://doi.org/10.1136/BMJ.F2539

Lee DTS, Sahota D, Leung TN, Yip ASK, Lee FFY, Chung TKH (2006) Psychological responses of pregnant women to an infectious outbreak: a case-control study of the 2003 SARS outbreak in Hong Kong. J Psychosom Res 61(5):707–713. https://doi.org/10.1016/j.jpsychores.2006.08.005

Li Q, Loke AY (2013) A spectrum of hidden morbidities among spousal caregivers for patients with cancer, and differences between the genders: a review of the literature. Eur J Oncol Nurs 17(5):578–587. https://doi.org/10.1016/j.ejon.2013.01.007

Lohm D, Flowers P, Stephenson N, Waller E, Davis MD (2014) Biography, pandemic time and risk: pregnant women reflecting on their experiences of the 2009 influenza pandemic. Health 18(5):493–508. https://doi.org/10.1177/1363459313516135

Lund C (2014) Poverty and mental health: towards a research agenda for low and middle-income countries. Commentary on Tampubolon and Hanandita (2014). Soc Sci Med 111:134–136. https://doi.org/10.1016/j.socscimed.2014.04.010

Marie Knaul F, Farmer PE, Krakauer EL, De Lima L, Bhadelia A, Jiang Kwete X et al (2018) The lancet commissions alleviating the access abyss in palliative care and pain relief-an imperative of universal health coverage: the lancet commission report. Lancet 391:1391–1454. https://doi.org/10.1016/S0140-6736(17)32513-8

McClarty L, Lazarus L, Pavlova D, Reza-Paul S, Balakireva O, Kimani J et al (2022) Socioeconomic burdens of the COVID-19 pandemic on LMIC populations with increased HIV vulnerabilities. Curr HIV/AIDS Rep 19:76–85. https://doi.org/10.1007/s11904-021-00591-w

Mousley E, Deribea K, Tamiru A, Tomczyk S, Hanlon C, Davey G (2014) Mental distress and podoconiosis in northern Ethiopia: a comparative cross-sectional study. Int Health. https://doi.org/10.1093/inthealth/ihu043

Mpango RS, Ssembajjwe W, Rukundo GZ, Birungi C, Kalungi A, Gadow KD et al (2023) Physical and psychiatric comorbidities among patients with severe mental illness as seen in Uganda. Eur Arch Psychiatry Clin Neurosci 273(3):613. https://doi.org/10.1007/S00406-022-01478-6

Mukherjee S (2020) Disparities, desperation, and divisiveness: coping with COVID-19 in India. Psychol Trauma Theory Res Pract Policy 12(6):582. https://doi.org/10.1037/TRA0000682

Musyimi CW, Mutiso VN, Nyamai DN, Ebuenyi I, Ndetei DM (2020) Suicidal behavior risks during adolescent pregnancy in a low-resource setting: a qualitative study. PLoS One 15(7):e0236269. https://doi.org/10.1371/journal.pone.0236269

Ng J, Sham A, Tang PL, Fung S (2004) SARS: pregnant women's fears and perceptions. Br J Midwifery 12(11):698–702. https://doi.org/10.12968/bjom.2004.12.11.16710

Punamäki RL, Palosaari E, Diab M, Peltonen K, Qouta SR (2015) Trajectories of posttraumatic stress symptoms (PTSS) after major war among Palestinian children: trauma, family- and child-related predictors. J Affect Disord 172:133–140. https://doi.org/10.1016/j.jad.2014.09.021

Purgato M, Gastaldon C, Papola D, van Ommeren M, Barbui C, Tol WA (2018) Psychological therapies for the treatment of mental disorders in low- and middle-income countries affected by humanitarian crises. Cochrane Database Syst Rev 2018(7):CD011849. https://doi.org/10.1002/14651858.CD011849.pub2

Rathod S, Pinninti N, Irfan M, Gorczynski P, Rathod P, Gega L, Naeem F (2017) Mental health service provision in low- and middle-income countries. Health Serv Insights 10:117863291769435. https://doi.org/10.1177/1178632917694350

Ross CA, Goldner EM (2009) Stigma, negative attitudes and discrimination towards mental illness within the nursing profession: a review of the literature. J Psychiatr Ment Health Nurs 16(6):558–567. https://doi.org/10.1111/j.1365-2850.2009.01399.x

Sabes-Figuera R, McCrone P, Bogic M, Ajdukovic D, Franciskovic T, Colombini N et al (2012) Long-term impact of war on healthcare costs: an eight-country study. PLoS One 7(1):e29603. https://doi.org/10.1371/journal.pone.0029603

Sepúlveda-Loyola W, Rodríguez-Sánchez I, Pérez-Rodríguez P, Ganz F, Torralba R, Oliveira DV, Rodríguez-Mañas L (2020) Impact of social isolation due to COVID-19 on health in older people: mental and physical effects and recommendations. J Nutr Health Aging 24(9):938–947. https://doi.org/10.1007/s12603-020-1469-2

Slone M, Lavi I, Ozer EJ, Pollak A (2017) The Israeli-Palestinian conflict: meta-analysis of exposure and outcome relations for children of the region. Child Youth Serv Rev 74:50–61. https://doi.org/10.1016/j.childyouth.2017.01.019

Smith DJ, Langan J, McLean G, Guthrie B, Mercer SW (2013) Schizophrenia is associated with excess multiple physical-health comorbidities but low levels of recorded cardiovascular disease in primary care: cross-sectional study. BMJ Open 3(4):e002808. https://doi.org/10.1136/BMJOPEN-2013-002808

Sprecher A, Van Herp M, Rollin PE (2017) Clinical management of Ebola virus disease patients in low-resource settings. Curr Top Microbiol Immunol 411:93–113. https://doi.org/10.1007/82_2017_18

Stranix-Chibanda L, Chibanda D, Chingono A, Montgomery E, Wells J, Maldonado Y et al (2005) Screening for psychological morbidity in HIV-infected and HIV-uninfected pregnant women using community counselors in Zimbabwe. J Int Assoc Phys AIDS Care 4(4):83–88. https://doi.org/10.1177/1545109706286555

Tee M, Wang C, Tee C, Pan R, Reyes PW, Wan X et al (2021) Impact of the COVID-19 pandemic on physical and mental health in lower and upper middle-income Asian countries: a comparison between the Philippines and China. Front Psych 11:568929. https://doi.org/10.3389/fpsyt.2020.568929

The World Bank. (n.d.) Population, total – lower middle income data. https://data.worldbank.org/indicator/SP.POP.TOTL?locations=XN. Accessed 4 Mar 2023

Tol WA, Murray SM, Lund C, Bolton P, Murray LK, Davies T et al (2019) Can mental health treatments help prevent or reduce intimate partner violence in low- and middle-income countries? A systematic review. BMC Womens Health 19(1):34. https://doi.org/10.1186/s12905-019-0728-z

Tora A, Mengiste A, Davey G, Semrau M (2018) Community involvement in the care of persons affected by podoconiosis—a lesson for other skin NTDs. Trop Med Infect Dis 3(3):87. https://doi.org/10.3390/tropicalmed3030087

Trautmann S, Rehm J, Wittchen H (2016) The economic costs of mental disorders. EMBO Rep 17(9):1245–1249. https://doi.org/10.15252/embr.201642951

Trivedi D (2017) Cochrane review summary: interventions for improving outcomes in patients with multimorbidity in primary care and community settings. Prim Health Care Res Dev 18(2):109–111. https://doi.org/10.1017/S1463423616000426

Veldhuis CB, Nesoff ED, McKowen ALW, Rice DR, Ghoneima H, Wootton AR et al (2021) Addressing the critical need for long-term mental health data during the COVID-19 pandemic: changes in mental health from April to September 2020. Prev Med 146:106465. https://doi.org/10.1016/j.ypmed.2021.106465

Vigo D, Thornicroft G, Atun R (2016) Estimating the true global burden of mental illness. Lancet Psychiatry 3(2):171–178. https://doi.org/10.1016/S2215-0366(15)00505-2

Vistorte AOR, Ribeiro WS, Jaen D, Jorge MR, Evans-Lacko S, Mari J d J (2018) Stigmatizing attitudes of primary care professionals towards people with mental disorders: a systematic review. Int J Psychiatry Med 53(4):317–338. https://doi.org/10.1177/0091217418778620

Wang C, Tee M, Roy AE, Fardin MA, Srichokchatchawan W, Habib HA et al (2021) The impact of COVID-19 pandemic on physical and mental health of Asians: a study of seven middle-income countries in Asia. PLoS One 16(2):e0246824. https://doi.org/10.1371/journal.pone.0246824

Wessells MG (2009) Do no harm: toward contextually appropriate psychosocial support in international emergencies. Am Psychol 64(8):842–854. https://doi.org/10.1037/0003-066X.64.8.842

WHO Health Workforce Department. (2021) The impact of COVID-19 on health and care workers: a closer look at deaths. (September)

Zakham F, Alaloui A, Levanov L, Vapalahti O (2019) Viral haemorrhagic fevers in the middle East. Rev Sci Tech 38(1):185–198. https://doi.org/10.20506/rst.38.1.2952

Strategies for Sustainable Preparedness and Response in LMIC

8

Abstract

Low-and Middle-income countries are faced with various situations some expected while others are unexpected. Among the expected, are chronic endemic diseases as well as seasonal outbreaks as well as unexpected disease outbreaks including emerging and re-emerging diseases.

While endemic diseases are expected, a certain level of preparedness is required to strategize on appropriate responses to prevent, control and or treat the disease. Economically, LMIC have varying health system and financial capacities to manage emergencies and crises. While some can withstand short term shocks, other countries can be overwhelmed with gross repercussions. Sustainable recovery, aligned with the SDG, requires cross-sectoral actions and mechanisms to manage unavoidable trade-offs between short and long-term priorities, and between economic, social, and environmental policy goals. Strategic planning and community awareness are crucial in managing health emergencies and building resilience. The "Sinkhole Phenomenon" is a reality in LMIC, the more the challenges in LMIC are overlooked and or ignored, the weaker the resilience in LMIC. Thus, strategic planning is an area worth investing in, by setting frameworks that are in line with IHR; as well as inclusive to meet the needs of the affected populations in LMIC, according to prevailing culture, norms, resources, and setting. There is hope of better outcomes for sustainable health if LMIC can build strong health systems.

Keywords

Low- and Middle-Income Countries (LMIC) · Preparedness and response · Emerging and re-emerging diseases

8.1 Preparedness and Response in Low and Middle Income Countries (LMIC)

Low-and Middle-income countries are faced with various situations some are expected while others are unexpected. Among the expected are chronic endemic diseases as well as seasonal outbreaks as well as unexpected disease outbreaks including emerging and re-emerging diseases.

While endemic diseases are expected, a certain level of preparedness is required to strategize on appropriate responses to prevent, control and or treat the disease.

As part of preparedness various factors are to be considered, these include human factors (e.g., population density, health, and wellbeing of the population), the environment (including proximity to reservoirs of disease, exposure to disease

F. A. Mphande, *Sustainable Health in Low and Middle Income Countries*, Sustainable Development Goals Series, https://doi.org/10.1007/978-981-99-4254-1_8

causing pathogens, vector distribution and climate), economic capabilities and infrastructure (including strength of the health system, availability of resources to manage the disease, manpower as well as skills to manage the disease. Economically, LMIC have varying health systems and financial capacities to manage emergencies and crises. While some health systems can withstand short term shocks, others can be overwhelmed with gross repercussions.

Preparedness therefore requires that countries have actual knowledge of what they are preparing for. To set up an effective strategy, there is need for prior knowledge and understanding of the problem at hand. The knowledge could be from previous experiences, data from other sources with a comparable situations, as well as access to information. Similarly, an effective response, requires ample resources to achieve implement and achieve the preparedness strategy. Financial resources are critical in achieving better preparedness and response. With a financially stable country, various response strategies can be explored and achieved. On the contrary, in countries with limited resources, they can only prepare as far as the available resources allow.

A well prepared strategy could assist in eliciting a better response. The COVID-19 pandemic has revealed gaps in countries and on a global scale how ill prepared countries were for the pandemic of such magnitude (Caballero-Anthony 2021; Chua et al. 2020; Docherty et al. 2020; Rahman 2020; Tusabe et al. 2022). What are some of the challenges that were observed during the COVID-19 pandemic and how do these experiences shape the future of preparedness and response? How can LMIC achieve sustainable preparedness and response?

8.1.1 Vulnerabilities in Preparedness and Universal Health Coverage (SDG3.8)

Geographically, LMIC vary, in setting, biodiversity, environments and governance. With varying systems of governance, each country and or region may have unique vulnerabilities most of which are rooted in inequality including weak governance, discrimination, injustice (Chua et al. 2020; Domapielle et al. 2022; Hasan et al. 2021; Rosa et al. 2022). Weak governance as well as population growth has led to degradation of biodiversity which could have serious implications to climate as well as the spread (transmission) of infectious disease (Domapielle et al. 2022; Gilson et al. 2017; Rosa et al. 2022). One challenge that has been reported for LMIC is the limited and or lack of infrastructure and services to curter for their respective populations (Domapielle et al. 2022; Eccleston-Turner and Upton 2021; Hasan et al. 2021). These limited capacities have affected preparedness and response in LMIC. Sustainable development goal 3.8 (SDG3.8), universal health coverage, is crucial and to attain this, LMIC should be better prepared for any form of emergency. Building resilient infrastructure (SDG3d) in both urban and rural communities of LMIC would assist in better preparedness and response (Javadi et al. 2020). It has been observed that while urban areas of LMIC may be able to provide health services, rural communities are often left out with limited or no health services (Akseer et al. 2020; Hartley 2004; Mishra et al. 2021; Monroe et al. 1992; Smith et al. 2008; Wichmann and Wichmann 2022). These inequalities affect the quality of healthcare in rural populations which are often exposed to emerging and re-emerging disease threats due to their livelihoods as well as NCD which often go unrecognized and poorly managed (Bocquier et al. 2014; Maman et al. 2018; McCollum and Damon 2014; Mediannikov et al. n.d.; Peer 2015; Ryll et al. 2018; Zafar and McQueen 2011).In the urban setting, disparities between the rich and poor urban populations have been reported (Aberese-Ako Id et al. 2022; Das et al. 2018; Mubarak et al. 2016; Nuwematsiko et al. 2022; Terefe et al. 2017). Disease distribution, access to healthcare services, quality of life and standards of living vary considerably, with populations in slums disproportionately affected by diarrhoeal diseases as well as vector borne diseases such as malaria, dengue, skin infections and respiratory diseases such as TB and COVID-19 (Brotherhood et al.

2022; Das et al. 2018; Isaakidis et al. 2011; Moto et al. 2015; Schuster et al. 2011; Snyder et al. 2016). Another vulnerable population is that of indigenous populations and hard to reach populations. Some indigenous communities live in isolated regions of the world far from public transportation, health services and any social and financial services. This isolation also implies limited contact with individuals from other communities. These populations are vulnerable to various infections including emerging and re-emerging diseases. Disparities in access to COVID-19 vaccines was amplified in indigenous communities most of which had to travel long distances to get treatment but also faced discrimination during the process (Agoramoorthy and Hsu 2022; Argoty-Pantoja et al. 2021; Huyser et al. 2022; Machado et al. 2022). Understanding vulnerabilities in various populations could assist in identifying gaps and creating solutions that are suitable to meet the healthcare needs for such populations. Universal health coverage for all means reaching populations in all corners of the world despite their social and economic status.

8.1.2 Investing in Research and Building Capacity

Building resilient infrastructure in LMIC is the focus of SDG3d. Such infrastructure include sustainable surveillance systems, access to primary healthcare, skilled health professionals as well as strong health systems that can withstand all forms of emergencies. Using the COVID-19 pandemic as an example, most LMIC and countries globally were ill prepared for a crisis of such magnitude. The infrastructure, skill set, equipment, and financial resources available could not match the response demands for COVID-19. The number of people infected was so enormous that hospitals and health personnel were overwhelmed. The infrastructure in place was not enough to manage the ever-increasing number of infections (OECD 2022). The mortality rate was so high causing distress on healthcare personnel and the population at large (Arshad Ali et al. 2020). The mental health toll on healthcare professionals was so enormous more so on the communities most of whom did not have access to accurate information on the pandemic, its origins, mode of transmission and health outcomes after infection (Hall 2020; Mukherjee 2020; Tasnim et al. 2020; Veldhuis et al. 2021). While high income countries were able to acquire and or manufacture most of the required equipment such as PPE, ventilators, as well as increasing research capacity, LMIC were limited on their responses due to constrained finances and skills capacity as well as infrastructure (Jowell and Barry 2020; McClarty et al. 2022; Rahman 2020). While rapid molecular testing and sequencing was available to determine and distinguish the various COVID-19 strains, in HIC, most LMIC did not have the capacity to do so (Khankeh et al. 2021; Sarfraz et al. 2022; Tran et al. 2020). Rapid diagnostic tools were designed and manufactured and were on the market in a short period of time, thanks to the research capacity in HIC and some middle-income countries (MIC) (Drain 2022). COVID-19 vaccines were available within the first year of the pandemic and were approved and rolled out for use globally. Although this was possible in HICs, LMIC were far behind to catch up on these advances due to limited capacity in infrastructure, finances, intellectual property, as well as skills. The pandemic revealed how deep the inequality gap is between HIC and LMIC and how the gap could affect the level of preparedness and response to an emergency (Dionne and Turkmen 2020; Khankeh et al. 2021). Research data was able to reveal patterns that were useful in strategizing and management of preparedness and response plans (WHO Health Workforce Department 2021). Building better preparedness and response infrastructure in LMIC is essential if universal health coverage is to be achieved globally. The fact that LMIC are behind on innovation and technology, skills and infrastructure, it is difficult to implement some of the existing technologies in these countries. Strengthening capacity in LMIC through better training, sustainable surveillance systems that can generate real time data as well as financial resources to acquire PPE, therapeutics and necessary vaccines could prove essential. Encouraging manufacture of materials such as PPE, masks, diagnostic kits, vaccines and other resources that are required for routine treatment of endemic diseases in LMIC would be critical. Multisectoral, national and international collaborations between various sectors would assist in strengthening capacity of LMIC in managing and combating health emergencies.

8.1.3 Digital Transformation

Digital technology has allowed for advancement in various fields of practice, including healthcare. From real-time data collection, diagnostics and analysis, the power of digital technology has transformed healthcare systems globally. These advancements have not been distributed equally, while some LMIC are able to join the digital transformation era, others still remain behind. Digital transformation requires various resources including human factors, technical factors, and healthcare ecosystem (Labrique et al. 2018).

While other countries are advancing, LMIC remain on the side lines with limited finances to acquire, maintain and upgrade their healthcare systems, limited skills to manage equipment and poor unsustainable governance. LMIC are yet to catch up with the digital transformation to improve the health of their populations. From telemedicine to data acquisition and sharing, access to health information and digital platforms for analysis and interpretation, LMIC have a long way to go to attain resilient infrastructure (SDG3d) as well as universal health coverage (SDG3.8)(Al-Shamsi et al. 2020; Denburg et al. 2019; Palagyi et al. 2019; Rao and Lombardi 2009). Digital transformation requires connectivity and constant power supply. Challenges in internet access and power supply in both urban and rural populations of LMIC is a challenge in attaining digital connectivity between institutions as well as community. While a telephone call can easily remind a patient of their hospital appointment, a video call used to ensure that a patient with TB is taking their medication on time, as well as an online consultation with a doctor, these are possible with internet access (Edwards and Patel 2003; Neubeck et al. 2009; Wu et al. 2015). Telemedicine has been used for communicable and non-communicable diseases with favourable outcomes.

Wearable devices are gaining traction globally. These devices are useful in monitoring, collection of real time data but also as early warning systems for the individuals of the impending health risk (Sharma et al. 2021). Wearable devices are crucial for monitoring, diagnostics and treatment of various conditions including chronic disease, cardiovascular disease and other neurological diseases (Barteit et al. 2021; Garbern et al. 2019; Sharma et al. 2021). Although these devices are available most often in HIC, the cost is quite high for the majority populations in LMIC. Despite the cost, some countries in LMIC are incorporating some wearable devices in their primary healthcare systems (Garbern et al. 2019; Jobarteh et al. 2020). Wearable could assist in alleviating the data gap and provide information that could assist in early mitigation of conditions in LMIC (Barteit et al. 2021; Jobarteh et al. 2020; Sharma et al. 2021; Zhang et al. 2022). Wearable devices are an alternative which could assist in monitoring and mitigating health risks in LMIC, these technologies could be prove useful in monitoring epidemic prone as well as emerging and re-emerging diseases in high risk populations globally.

8.1.4 Using the Past to Build a Better Future

Economic vulnerabilities: Despite COVID-19 being a global pandemic, each country experienced the pandemic rather uniquely. The transmission rate, mortality rates, number of infections and recoveries varied between countries. Emerging economies suffered the scourge of the pandemic due to pre-existing challenges such as existing epidemic diseases, including HIV/AIDS, malaria, TB, neglected tropical diseases and other public health challenges (Priorities 2022). Apart from the universal global impacts such as disruption of travel and tourism, loss of businesses, the economic downturn and social impacts; some of the impacts were specific to countries and or regions (https://blogs.worldbank.org/voices/2020-year-review-impact-covid-19-12-charts). There was an increase in inequalities as the gap between the rich and the poor grew due to economic losses among vulnerable populations most of whom their jobs were greatly affected by lockdowns, as well as the slowing down of informal business sector (Bundervoet et al. 2021). The pandemic has shown the fragility of economies in LMIC as well as the growth in inequalities due to differences in resilience between rich and poor populations within the same country as well as globally. Designing policies that could support vulnerable communities by protecting their economic well-being while safeguarding health could assist in reducing inequality (Adjorlolo and Egbenya 2020). The current strategies work in favour of those who are better off, as opposed to those struggling economically (Adjorlolo and Egbenya 2020; Lau et al. 2020).

Use of strategies that are culturally appropriate and conducive for the population and setting: The current SDG show the fragility of LMIC from all angles; economy, capacity, infrastructure, skill shortages, risk, epidemic, research and development, affordable medicines and vaccines, health financing, recruitment and training, risk reduction and management, early warning, and management of global health risks (see Quote 8.1). With such a list of vulnerabilities it is daunting to fathom that with such differences the same set of policies that work in HIC should be expected to work in LMIC. The task faced by LMIC to achieve SDG3 and its targets is huge as such strategies should be put in place that would support LMIC to achieve the targets to support the needs within the countries. Identifying the current needs and addressing those needs while working towards sustainability at a global scale is possible but requires global commitment and acknowledgement that the scales are tipped and unless these scales are balanced, sustainability is out of question (Fig. 8.1). The current global inequalities in preparedness and response bring the question, what is it that is being sustained? The status quo, where LMIC remain vulnerable or universal equity and equality?

It is worth noting that as health emergencies appear and disappear, the base line for LMIC is quite different from that of HIC (Fig. 8.1). When a health emergency strikes in LMIC, the environment is already plagued with endemic infectious diseases such as malara, TB, cholera, helminthiasis, neglected tropical diseases, and HIV/AIDS which already strain the health systems (Ampah et al. 2016; Assefa and Gilks 2020; Hunsperger et al. 2019; Jones et al. 2008; Knopp et al. 2010; Msyamboza et al. 2012; Seimenis 2012; Sutton 2018; La Vincente et al. 2009). With limited resources, adding on a health emergency of the magnitude of for example Ebola, and COVID-19 could paralyze the system with repercussions on the existing endemic diseases (Bundervoet et al. 2021; Fagerlin et al. 2017). As such preparedness and response policies in LMIC should take into consideration the existing disease burden in a particular setting. Quote 8.1 points out the various targets of SDG3, which are crucial if LMIC are to attain SDG3. Various programmes are already in place for the epidemics of HIV, malaria, TB, NTD, hepatitis, water-borne diseases and other communicable diseases (Elphick-Pooley and Engels 2020; Engels and Zhou 2020; Hsü and Hsü 2017; UNAIDS 2016; World Health Organization 2020a). These programmes are targeted to reduce the burden of the disease, prevent and control the spread of the disease but also eliminate the diseases (Julia Nicole Simac et al. 2017; Klepac et al. 2013; WHO 2015).

Quote 8.1

Target 3.3: *By 2030, end the epidemics of AIDS, tuberculosis, malaria and neglected tropical diseases and combat hepatitis, water-borne diseases, and other communicable diseases.*

Target 3.B *Support the research and development of vaccines and medicines for the communicable and noncommunicable diseases that primarily affect developing countries, provide access to affordable essential medicines and vaccines, in accordance with the Doha Declaration on the TRIPS Agreement and Public Health, which affirms the right of developing countries to*

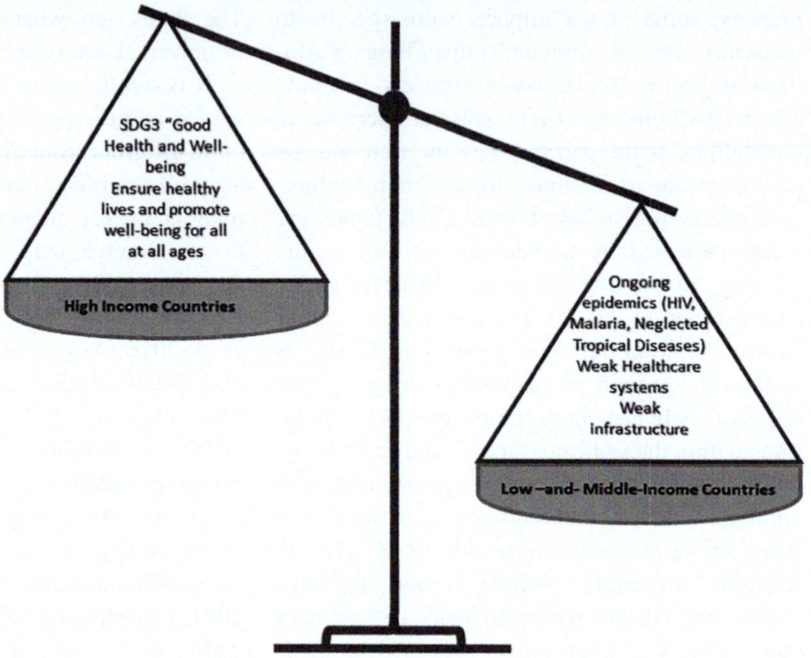

Fig. 8.1 SDG3 in high and low-middle income countries, LMIC carry a huge public health burden which has thrown SDG3 off balance

use to the full the provisions in the Agreement on Trade Related Aspects of Intellectual Property Rights regarding flexibilities to protect public health, and, in particular, provide access to medicines for all.

Target 3.C *Substantially increase health financing and the recruitment, development, training, and retention of the health workforce in developing countries, especially in least developed countries and small island developing States.*

Target 3.D *Strengthen the capacity of all countries, in particular developing countries, for early warning, risk reduction and management of national and global health risks.*

8.2 Strategic Planning and Anticipating Future Risks

Adaptive policies: By understanding the weaknesses and strengths of the affected countries' health systems, geographical and predisposed environmental/climatic changes, as well as cultural norms and values, it is possible to create adaptive strategies.

Although this method may not be challenge proof, but it is a good start. Every country should be able to adapt policies according to their setting and populations. By assessing the successes and challenges from the existing policies, countries should be able to build on their experiences.

Risk assessments, early warning systems are still weak in most LMIC, this is an area that has to be strengthened. Countries need to be more proactive than reactive to disease outbreaks and health emergencies. There is need to set up frameworks of preparedness and response for various situations including all forms of outbreaks. The WHO has published frameworks on various infectious diseases, these frameworks could be adapted to the various settings and countries in LMIC could benefit in the long run (Bergeri et al. 2022; Calisher et al. 2019; Engels and Zhou 2020; Gilson et al. 2017; Kochhar et al. 2019). For example countries should be able to have frameworks on how to prepare and respond to outbreaks from waterborne pathogens such as cholera, airborne viral and or bacterial pathogens such as influenza, meningitis, COVID-19, as well as fungal and zoonotic pathogens. By understanding priority pathogens, the mode of transmissions of endemic and unknown pathogens, the populations at risk, pre-existing risk factors and the possible response strategies according to available resources is crucial (Chatterjee et al. 2021; Lamorde et al. 2018; Wijesinghe et al. 2020; World Health Organization 2020b). It has been reported that LMIC are struggling with infrastructure, limited capacity as well as limited resources such as medicines and vaccines, thus investing in better and stronger prevention and control strategies could assist in keeping certain diseases at bay. Similarly, having a solid preparedness and response framework could assist in the management and control of an outbreak, thereby reducing the spread of the disease (Jourdain et al. 2019; Pigott et al. 2017; Rampal et al. 2020; Wignjadiputro et al. 2020; Ziegler et al. 2018).

Community awareness is another important aspect that has to be considered as part of the preparedness and response plan. While governments and health personnel can put together frameworks and strategies, the communities on whom these frameworks and strategies are meant for should be made aware beforehand. Thus involving communities at the grassroots in the preparation of preparedness and response strategy is an essential component. If communities are aware of what is required in the likely event of a certain disease outbreak, they would be in the right position to respond positively than being bombarded by information during the outbreak. Lack of information on prevention strategies has resulted in misinformation where populations are not aware as to why they have to follow certain protocols and or change their norms in order to combat the disease. If populations are aware of what needs to be done to prevent and control the spread of infections, they are likely to be more prepared both physically and psychologically on what is expected and their role in the process (Ackumey et al. 2011; Barogui et al. 2018; Cantey et al. 2013; De Schacht et al. 2019; Kanu et al. 2014; Rimi et al. 2016; SteelFisher et al. 2015).

Working with communities could assist in identifying unseen challenges and opportunities, designing innovative and fitting policy actions and strategies, and help stakeholders have a better perspective on the challenges at the grassroots (De Schacht et al. 2019; Khan and Qazi 2013; Kung et al. 2015; Msyamboza et al. 2012; Mushtaq et al. 2015; SteelFisher et al. 2015). Using pre-existing and or regular disease outbreaks and presenting scenarios of possible outbreaks from various pathogens could assist in informing the communities of future challenges as well as allow the communities in planning for these challenges. Training communities to have a different mindset and perspective on their role in public health, health emergencies and disease prevention and control is crucial. Changing the mindset of helplessness to a more self-sustaining and competent attitude could assist in building stronger and resilient communities able to tackle various challenges than always waiting for a saviour to help them.

8.3 Resilience and the Sinkhole Phenomenon

The Oxford dictionary defines resilience as “the capacity to withstand or to recover quickly from difficulties; toughness” https://www.google.com/search?q=resilience&oq=resilience&aqs=chrome..69i57j0i131i433i512j0i512l7j46i512.2302j0j15&sourceid=chrome&ie=UTF-8 .

From the definition above, three keywords stand out; “capacity”, “withstand”, “recover quickly”. Low- and middle-income countries are inundated with various challenges including governance, disease outbreaks, neglected tropical diseases, endemic tropical diseases, poorly functioning infrastructure, as well as limited capacity (Fig. 8.2).

These challenges have been exacerbated by the increasing gaps in public healthcare and public health systems that allow for “internal erosion”. The destruction caused by these gaps is not immediately visible but slowly eats away the struggling healthcare and public health foundations which culminates in the crumbling of resilience. For example, poor governance in LMIC has affected health systems development and

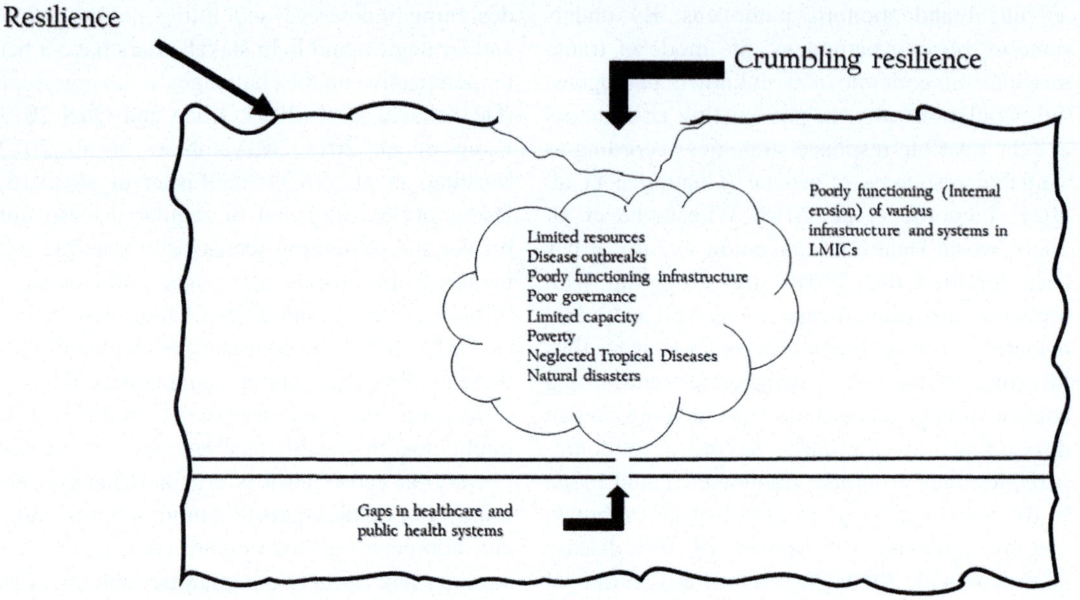

Fig. 8.2 The sinkhole phenomenon, existing gaps in healthcare systems in LMIC create holes which if left unattended will affect resilience in these regions

advancements in most of these countries (Chua et al. 2020; Denburg et al. 2019; Domapielle et al. 2022; Gilson et al. 2017; Kirigia and Barry 2008; Mushtaq et al. 2015). Corruption, political instability, war and displacement of populations has negatively affected development of societies. Misappropriation of funds in some LMIC have resulted in poor health research development and negative health outcomes in the affected populations (Fielding 2010; Kirigia and Barry 2008; Zafar and McQueen 2011; Zhang et al. 2014). Vaccine coverage and distribution has been greatly affected in countries facing conflict, poorly affecting children immunization as well as routine immunization for various vaccine preventable diseases (Bockarie et al. 2013; Burkle Jr 2017; Haq et al. 2013; Njoh 2002; Sabes-Figuera et al. 2012). Since these internal challenges are often dealt with separately, "patching" one of the gaps may result in temporary resolution of the issue in the short term but may not have lasting solutions. In the end, holes develop in resilience, forcing the country to tumble into the resulting sinkhole. The sinkhole phenomenon could assist in visualizing the challenges in LMIC and create better and innovative ways to manage the healthcare and public health situation in LMIC. So what is the concept of the sinkhole phenomenon? Naturally a sinkhole is caused by faster and more turbulent underground water that is more erosive and eats away the softer parts of natural bedrock, as more and more rock and soil is eroded, a big gaping hole develops and swallows everything in its way. Applying this phenomenon to public health in LMIC, there are many challenges that put pressure on health systems in LMIC, these affect their preparedness and response strategies and in the end these countries are plunged deeper into health situations (sinkhole) with minimal chances of recovery. While early warning systems might be available, not all countries have the capacity to prepare for the challenges faced. The limited capacity whether in resources, infrastructure, skill sets, finances and more create gaping holes that are difficult to fix. Thus critical decision have to be made by LMIC in identifying innovative and sustainable ways to manage the various situations faced by these countries. Low- and middle-income countries are currently facing natural disasters related climate change, which further affect the already existing health, economic, and social crises faced in this region. The

COVID19 pandemic has further complicated the situation (Abu-Odah et al. 2022; Deml and Githaiga 2022; Tagoe et al. 2021; Troiano and Nardi 2020).

The Organisation for Economic Co-operation and Development (OECD) developed a Policy Coherence for Sustainable Development (PCSD) to help countries within the OECD to meet SDG (OECD 2022). A summary of the policies is shown in the table (Table 8.1).

The policies in Table 8.1 could be applied to many countries globally to achieve the SDG. With differences existing between countries, the decisions taken will depend on the existing infrastructure, capacities and context/setting (OECD 2022).

Table 8.1 OECD policies and goals to achieve SDG (*source* OECD 2022)

Policy	Goal
Build political commitment and leadership at the highest level	To shape the national debate and support recovery strategies aligned with the SDG, and to mobilize whole-of-government action and orient recovery measures towards sustainable development
Adopt a strategic long-term vision	To make informed choices and considering the consequences of today's decisions and building resilience against the long-term shocks that will affect the economy, society, and environment
Strengthen policy integration	To balancing often divergent economic, social, and environmental priorities for recovery
Ensure whole-of-government coordination	To identify and address policy divergences and conflicts between measures for recovery and achievement of the SDG
Engage subnational levels of government appropriately	To align priorities and promote coordinated action at various levels of government for sustainable recovery
Engage key stakeholder effectively	To ensure that different voices are heard and work with all relevant actors to identify challenges, set priorities, align actions, and mobilize resources for recovery
Analyse and assess policy impacts	To consider the effects of domestic recovery measures on global sustainable development
Strengthen monitoring, reporting and evaluation systems	To inform the design of coherent and sustainable recovery strategies and adjust recovery measures in light of potential negative effects on sustainable development

8.4 Regional Cooperation

In LMIC the various constraints often affect the speed of response, extent of response and longevity of response. Rapid case detection is crucial in identifying new cases and curbing the spread of the disease. Early detection and response means diseases can be managed before they spiral out of control, before they spread further and possibly before onset of complications of disease (Aceng et al. 2020; Esser et al. 2019; Msyamboza et al. 2012). Rapid detection owes to availability of effective surveillance and availability of laboratories, skills, and infrastructure to identify disease causing pathogens. Most LMIC have limited laboratory capacity to run even basic laboratory tests including serological and molecular diagnostics (Diarra et al. 2016; Osadebe et al. 2017; Stern et al. 2016). For better preparedness and response LMIC require effective governance which should be able to provide effective coordination of public health emergencies and outbreaks. Most LMIC are dependent on support from donors to achieve their healthcare needs (Grépin et al. 2017; Kollie et al. 2020). LMIC should therefore strive to build a strong workforce with the resources available within their countries and or regional collaboration. Community health workers (CHW) are an essential part of primary healthcare in LMIC as these are the ones who interact communities. Community health workers are crucial in identification of diseases, care as well as follow up of patients within communities (Andrews et al. 2009; Barogui et al. 2018; Mohammed et al. 2012; Molla et al. 2012). Countries in LMIC should develop functional frameworks which incorporate (CHW) both as individual countries and or as clusters/region to prepare and respond to outbreaks. With a group effort, countries could pool resources and or share efforts such as training as well as laboratory capacity to assist with early detection and management of disease outbreaks (Lee et al. 2020; Nkengasong et al. 2017).

8.5 Conclusion

Low and middle income countries are inundated by a myriad of health challenges which have been further complicated by social, economic, and environmental challenges. Vulnerabilities including health disparities between urban and rural populations as well rich and poor communities are growing in LMIC and globally. By understanding and addressing these vulnerabilities in respective populations, countries could identify gaps and create solutions that are suitable for their healthcare needs. This would assist in building and setting up infrastructure that could support the health needs of populations in LMIC. While digital transformation is still in its infancy, wearable devices could assist in monitoring, early detection and management of health issues including chronic diseases. All these possible solutions require good governance and financial commitment from respective governments. Sustainable recovery, aligned with the SDG, requires cross-sectoral actions and mechanisms to manage unavoidable trade-offs between short and long-term priorities, and between economic, social, and environmental policy goals. Sustainable development goal 3d (SDG 3.d), which encourages "strengthened capacity of all countries, in particular developing countries, for early warning, risk reduction and management of national and global health risks". The COVID-19 pandemic revealed vulnerabilities in preparedness at different levels including at national and global level. Although successes were observed with response efforts globally, the response showed gaps in solidarity inclining more towards nationalistic agendas. As LMIC work towards strengthening research capacity, detection, diagnosis, treatment and prevention as well as monitoring of infections, collaboration and coordination between countries and or regions is crucial to achieve SDG3 and all its targets. With struggling economies, working as a group of countries could assist LMIC to achieve better and effective results than working individually. With advancements in digital transformation globally, LMIC have to work together to strengthen capacity of digital technology. Strategic planning and community awareness are crucial in managing health emergencies and building resilience. The "Sinkhole Phenomenon" is a reality in LMIC, the more the challenges in LMIC are overlooked and or ignored, the deeper LMIC resilience plunges. Thus strategic planning is an area worth investing in, by setting frameworks that are in line with IHR as well as inclusive to meet the needs of the affected populations. The strategies should consider the prevailing culture, norms, resources and setting in LMIC, with such there is hope of better outcomes for sustainable health.

References

Aberese-Ako Id M, Dalaba MA, Anumu FEY, Ofosu A, Gyapong M (2022) The socio-economic and health effects of COVID-19 among rural and urban-slum dwellers in Ghana: a mixed methods approach. PLoS One 17(7):e0271551. https://doi.org/10.1371/journal.pone.0271551

Abu-Odah H, Su J, Musa SS (2022) Unwillingness or reluctance of Palestinians to get the COVID-19 vaccine: the reasons behind it and how to persuade them. Int J Infect Dis 119:53. https://doi.org/10.1016/J.IJID.2022.03.039

Aceng JR, Ario AR, Muruta AN, Makumbi I, Nanyunja M, Komakech I et al (2020) Uganda's experience in Ebola virus disease outbreak preparedness, 2018–2019. Glob Health 16(1):24. https://doi.org/10.1186/s12992-020-00548-5

Ackumey MM, Kwakye-Maclean C, Ampadu EO, de Savigny D, Weiss MG (2011) Health services for Buruli ulcer control: lessons from a field study in Ghana. PLoS Negl Trop Dis 5:e1187. https://doi.org/10.1371/journal.pntd.0001187

Adjorlolo S, Egbenya DL (2020) A twin disaster: addressing the COVID-19 pandemic and a cerebrospinal meningitis outbreak simultaneously in a low-resource country. Glob Health Action 13(1):1795963. https://doi.org/10.1080/16549716.2020.1795963

Agoramoorthy G, Hsu MJ (2022) COVID-19 and India's vulnerable indigenous populations. Am J Hum Biol 34(2):e23608. https://doi.org/10.1002/AJHB.23608

Akseer N, Kandru G, Keats EC, Bhutta ZA (2020) COVID-19 pandemic and mitigation strategies: implications for maternal and child health and nutrition. Am J Clin Nutr 112(2):251–256. https://doi.org/10.1093/ajcn/nqaa171

Al-Shamsi HO, Alhazzani W, Alhuraiji A, Coomes EA, Chemaly RF, Almuhanna M et al (2020) A practical approach to the management of cancer patients during the novel coronavirus disease 2019 (COVID-19) pandemic: an international collaborative group. Oncologist 25(6):e936–e945. https://doi.org/10.1634/theoncologist.2020-0213

Ampah KA, Asare P, Binnah DDG, Maccaulley S, Opare W, Röltgen K et al (2016) Burden and historical trend of Buruli ulcer prevalence in selected communities along the Offin River of Ghana. PLoS Negl Trop Dis 10(4):e0004603. https://doi.org/10.1371/journal.pntd.0004603

Andrews RM, Kearns T, Connors C, Parker C, Carville K, Currie BJ, Carapetis JR (2009) A regional initiative to reduce skin infections amongst aboriginal children living in remote communities of the Northern Territory, Australia. PLoS Negl Trop Dis 3(11):e554. https://doi.org/10.1371/journal.pntd.0000554

Argoty-Pantoja AD, Robles-Rivera K, Rivera-Paredez B, Salmerón J (2021) COVID-19 fatality in Mexico's indigenous populations. Public Health 193:69. https://doi.org/10.1016/J.PUHE.2021.01.023

Arshad Ali S, Baloch M, Ahmed N, Arshad Ali A, Iqbal A (2020) The outbreak of coronavirus disease 2019 (COVID-19)–an emerging global health threat. J Infect Public Health 13(4):644–646. https://doi.org/10.1016/j.jiph.2020.02.033

Assefa Y, Gilks CF (2020) Ending the epidemic of HIV/AIDS by 2030: will there be an endgame to HIV, or an endemic HIV requiring an integrated health systems response in many countries? Int J Infect Dis 100:273–277. https://doi.org/10.1016/J.IJID.2020.09.011

Barogui YT, Diez G, Anagonou E, Johnson RC, Gomido IC, Amoukpo H et al (2018) Integrated approach in the control and management of skin neglected tropical diseases in Lalo, Benin. PLoS Negl Trop Dis 12(6):e0006584. https://doi.org/10.1371/journal.pntd.0006584

Barteit S, Boudo V, Ouedraogo A, Zabré P, Ouremi L, Sié A et al (2021) Feasibility, acceptability and validation of wearable devices for climate change and health research in the low-resource contexts of Burkina Faso and Kenya: study protocol. PLoS One 16(9):e0257170. https://doi.org/10.1371/JOURNAL.PONE.0257170

Bergeri I, Lewis HC, Subissi L, Nardone A, Valenciano M, Cheng B et al (2022) Early epidemiological investigations: World Health Organization UNITY protocols provide a standardized and timely international investigation framework during the COVID-19 pandemic. Influenza Other Respir Viruses 16(1):7–13. https://doi.org/10.1111/irv.12915

Bockarie MJ, Kelly-Hope LA, Rebollo M, Molyneux DH (2013) Preventive chemotherapy as a strategy for elimination of neglected tropical parasitic diseases: endgame challenges. Philos Trans R Soc Lond Ser B Biol Sci 368(1623):20120144. https://doi.org/10.1098/rstb.2012.0144

Bocquier P, Collinson MA, Clark SJ, Gerritsen AAM, Kahn K, TollMan SM (2014) Ubiquitous burden: the contribution of migration to AIDS and tuberculosis mortality in rural South Africa. Etude Popul Afr 28(1):691–701. https://doi.org/10.11564/28-0-525

Brotherhood L, Cavalcanti T, Da Mata D, Santos C (2022) Slums and pandemics. J Dev Econ 157:102882. https://doi.org/10.1016/j.jdeveco.2022.102882

Bundervoet T, Dávalos ME, Garcia N (2021) The short-term impacts of COVID-19 on households in developing countries. World Bank Policy Research working paper 9582, (March)

Burkle FM Jr (2017) The politics of global public health in fragile states and ungoverned territories. PLoS Curr 9(Disasters). https://doi.org/10.1371/CURRENTS.DIS.BA3BEEDE71CA0746A0972AA3837ED618

Caballero-Anthony, M. (2021) COVID-19 in Southeast Asia: regional pandemic preparedness matters. Order from Chaos [BLOG].. http://libaccess.mcmaster.ca/login?url=https://www.proquest.com/blogs-podcasts-websites/covid-19-southeast-asia-regional-pandemic/docview/2477799720/se-2?accountid=12347%0Ahttp://sfx.scholarsportal.info/mcmaster?url_ver=Z39.88-2004&rft_val_fmt=info:ofi/fm

Calisher CH, Briese T, Brister JR, Charrel RN, Dürrwald R, Ebihara H et al (2019) Strengthening the interaction of the virology community with the international committee on taxonomy of viruses (ICTV) by linking virus names and their abbreviations to virus species. Syst Biol 68(5):828–839. https://doi.org/10.1093/sysbio/syy087

Cantey PT, Chuk MG, Kohl KS, Herrmann J, Weiss P, Graffunder CM et al (2013) Public health emergency preparedness: lessons learned about monitoring of interventions from the National Association of county and city health official's survey of nonpharmaceutical interventions for pandemic H1N1. J Public Health Manag Pract 19(1):70–76. https://doi.org/10.1097/PHH.0b013e31824d4666

Chatterjee P, Nair P, Chersich M, Terefe Y, Chauhan AS, Quesada F, Simpson G (2021) One health, "disease X" & the challenge of "unknown" unknowns. Indian J Med Res 153:264–271. https://doi.org/10.4103/ijmr.IJMR_601_21

Chua AQ, Tan MMJ, Verma M, Han EKL, Hsu LY, Cook AR et al (2020) Health system resilience in managing the COVID-19 pandemic: lessons from Singapore. BMJ Glob Health 5(9):e003317. https://doi.org/10.1136/bmjgh-2020-003317

Das M, Angeli F, Krumeich AJSM, Van Schayck OCP (2018) Patterns of illness disclosure among Indian slum dwellers: a qualitative study. BMC Int Health Hum Rights 18(1):3. https://doi.org/10.1186/s12914-018-0142-x

De Schacht C, Mutaquiha C, Faria F, Castro G, Manaca N, Manhiça I, Cowan J (2019) Barriers to access and adherence to tuberculosis services, as perceived by patients: a qualitative study in Mozambique. PLoS One 14(7):e0219470. https://doi.org/10.1371/journal.pone.0219470

Deml MJ, Githaiga JN (2022) Original research: determinants of COVID-19 vaccine hesitancy and uptake in sub-Saharan Africa: a scoping review. BMJ Open 12(11):e066615. https://doi.org/10.1136/BMJOPEN-2022-066615

Denburg AE, Ramirez A, Pavuluri S, McCann E, Shah S, Alcasabas T et al (2019) Political priority and pathways to scale-up of childhood cancer care in five nations. PLoS One 14(8):e0221292. https://doi.org/10.1371/JOURNAL.PONE.0221292

Diarra B, Safronetz D, Sarro YDS, Kone A, Sanogo M, Tounkara S et al (2016) Laboratory response to 2014 Ebola virus outbreak in Mali. J Infect Dis 214(Suppl.3):S164–S168. https://doi.org/10.1093/infdis/jiw200

Dionne KY, Turkmen FF (2020) The politics of pandemic othering: putting COVID-19 in global and historical context. Int Organ 74(S1):E213–E230. https://doi.org/10.1017/S0020818320000405

Docherty AB, Harrison EM, Green CA, Hardwick HE, Pius R, Norman L et al (2020) Features of 20 133 UK patients in hospital with covid-19 using the ISARIC WHO clinical characterisation protocol: prospective observational cohort study. BMJ 369:m1985. https://doi.org/10.1136/bmj.m1985

Domapielle MK, Sumankuuro J, Der Bebelleh F (2022) Revisiting the debate on health financing in low and middle-income countries: an integrative review of selected models. Int J Health Plann Manag 37(6):3061–3074. https://doi.org/10.1002/HPM.3566

Drain PK (2022) Rapid diagnostic testing for SARS-CoV-2. N Engl J Med 386(3):264–272. https://doi.org/10.1056/nejmcp2117115

Eccleston-Turner M, Upton H (2021) International collaboration to ensure equitable access to vaccines for COVID-19: the ACT-accelerator and the COVAX facility. Milbank Q 99(2):426–449. https://doi.org/10.1111/1468-0009.12503

Edwards MA, Patel AC (2003) Telemedicine in the state of Maine: a model for growth driven by rural needs. Telemed J E Health 9(1):25–39. https://doi.org/10.1089/153056203763317620

Elphick-Pooley T, Engels D (2020) World NTD Day 2022 and a new Kigali declaration to galvanise commitment to end neglected tropical diseases. Infect Dis Poverty 11(1):2. https://doi.org/10.1186/s40249-021-00932-2

Engels D, Zhou XN (2020) Neglected tropical diseases: an effective global response to local poverty-related disease priorities. Infect Dis Poverty 9(1):1–9. https://doi.org/10.1186/s40249-020-0630-9

Esser HJ, Mögling R, Cleton NB, van der Jeugd H, Sprong H, Stroo A et al (2019) Risk factors associated with sustained circulation of six zoonotic arboviruses: a systematic review for selection of surveillance sites in non-endemic areas. Parasit Vectors 12(1):265. https://doi.org/10.1186/s13071-019-3515-7

Fagerlin A, Valley TS, Scherer AM, Knaus M, Das E, Zikmund-Fisher BJ (2017) Communicating infectious disease prevalence through graphics: results from an international survey. Vaccine 35(32):4041–4047. https://doi.org/10.1016/J.VACCINE.2017.05.048

Fielding D (2010) Health aid and governance in developing countries. Health Econ 20:757–769. https://doi.org/10.1002/hec

Garbern SC, Mbanjumucyo G, Umuhoza C, Sharma VK, Mackey J, Tang O et al (2019) Validation of a wearable biosensor device for vital sign monitoring in septic emergency department patients in Rwanda. Digit Health 5:1–13. https://doi.org/10.1177/2055207619879349

Gilson L, Lehmann U, Schneider H (2017) Practicing governance towards equity in health systems: LMIC perspectives and experience Lucy Gilson. Int J Equity Health 16(1):1–5. https://doi.org/10.1186/s12939-017-0665-0

Grépin KA, Pinkstaff CB, Shroff ZC, Ghaffar A (2017) Donor funding health policy and systems research in low- and middle-income countries: how much, from where and to whom. Health Res Policy Syst 15(1):68. https://doi.org/10.1186/s12961-017-0224-6

Hall H (2020) The effect of the COVID-19 pandemic on healthcare workers' mental health. JAAPA 33(7):45–48. https://doi.org/10.1097/01.JAA.0000669772.78848.8c

Haq Z, Mahjour J, Khan W (2013) Communicable diseases in the eastern Mediterranean region: prevention and control 2010-2011. East Mediterr Health J 19(10):888–891

Hartley D (2004) Rural health disparities, population health, and rural culture. Am J Public Health 94(10):1675–1678

Hasan BS, Rasheed MA, Wahid A, Kumar RK, Zuhlke L (2021) Generating evidence from contextual clinical research in low- to middle income countries: a roadmap based on theory of change. Front Pediatr 9:764239. https://doi.org/10.3389/FPED.2021.764239

Hsü SY, Hsü HF (2017) Parasitic diseases in Africa and the western hemisphere: early documentation and tranmission by the slave trade. Am J Trop Med Hyg 19(5):894–895. https://doi.org/10.4269/ajtmh.1970.19.894

Hunsperger E, Juma B, Onyango C, Ochieng JB, Omballa V, Fields BS et al (2019) Building laboratory capacity to detect and characterize pathogens of public and global health security concern in Kenya. BMC Public Health 19(Suppl 3):477. https://doi.org/10.1186/s12889-019-6770-9

Huyser KR, Yellow Horse AJ, Collins KA, Fischer J, Jessome MG, Ronayne ET et al (2022) Understanding the associations among social vulnerabilities, indigenous peoples, and COVID-19 cases within Canadian health regions. Int J Environ Res Public Health 19(19):12409. https://doi.org/10.3390/IJERPH191912409

Isaakidis P, Cox HS, Varghese B, Montaldo C, da Silva E, Mansoor H et al (2011) Ambulatory multi-drug resistant tuberculosis treatment outcomes in a cohort of HIV-infected patients in a slum setting in Mumbai, India. PLoS One 6(12):1–9. https://doi.org/10.1371/journal.pone.0028066

Javadi D, Ssempebwa J, Isunju JB, Yevoo L, Amu A, Nabiwemba E et al (2020) Implementation research on sustainable electrification of rural primary care facilities in Ghana and Uganda. Health Policy Plan 35(Suppl. 2):II124–II136. https://doi.org/10.1093/HEAPOL/CZAA077

Jobarteh ML, McCrory MA, Lo B, Sun M, Sazonov E, Anderson AK et al (2020) Development and validation of an objective, passive dietary assessment method for estimating food and nutrient intake in households in lowand middle-income countries: a study protocol. Curr Dev Nutr 4(2):nzaa020. https://doi.org/10.1093/cdn/nzaa020

Jones KE, Patel NG, Levy MA, Storeygard A, Balk D, Gittleman JL, Daszak P (2008) Global trends in emerging infectious diseases. Nature 451(7181):990–993. https://doi.org/10.1038/nature06536

Jourdain F, Samy AM, Hamidi A, Bouattour A, Alten B, Faraj C et al (2019) Towards harmonisation of entomological surveillance in the Mediterranean area. PLoS Negl Trop Dis 13(6):e0007314. https://doi.org/10.1371/journal.pntd.0007314

Jowell A, Barry M (2020) COVID-19: a matter of planetary, not only national health. Am J Trop Med Hyg 103(1):31–32. https://doi.org/10.4269/ajtmh.20-0419

Julia Nicole Simac SB, Farber JA, Brako MYO, Giudice-Jimenez RAL, Raspa SS, Achore M, MacKnight SD (2017) Malaria elimination in Sri Lanka. J Health Spec 5:60–65. https://doi.org/10.4103/jhs.JHS

Kanu JS, Tang Y, Liu Y (2014) Assessment on the knowledge and reported practices of women on maternal and child health in rural Sierra Leone: a cross-sectional survey. PLoS One 9(8):e105936. https://doi.org/10.1371/journal.pone.0105936

Khan T, Qazi J (2013) Hurdles to the global antipolio campaign in Pakistan: an outline of the current status and future prospects to achieve a polio free world. J Epidemiol Community Health 67:696–702. https://doi.org/10.1136/jech-2012-202162

Khankeh H, Farrokhi M, Roudini J, Pourvakhshoori N, Ahmadi S, Abbasabadi-Arab M et al (2021) Challenges to manage pandemic of coronavirus disease (COVID-19) in Iran with a special situation: a qualitative multi-method study. BMC Public Health 21(1):1–9. https://doi.org/10.1186/s12889-021-11973-5

Kirigia JM, Barry SP (2008) Health challenges in Africa and the way forward. Int Arch Med 1:27. https://doi.org/10.1186/1755-7682-1-27

Klepac P, Metcalf CJE, McLean AR, Hampson K (2013) Towards the endgame and beyond: complexities and challenges for the elimination of infectious diseases. Philos Trans R Soc Lond Ser B Biol Sci 368(1623):20120137. https://doi.org/10.1098/rstb.2012.0137

Knopp S, Mohammed KA, Stothard JR, Khamis IS, Rollinson D, Marti H, Utzinger J (2010) Patterns and risk factors of helminthiasis and anemia in a rural and a peri-urban community in Zanzibar, in the context of helminth control programs. PLoS Negl Trop Dis 4(5):e681. https://doi.org/10.1371/journal.pntd.0000681

Kochhar S, Edwards KM, Ropero Alvarez AM, Moro PL, Ortiz JR (2019) Introduction of new vaccines for immunization in pregnancy – programmatic, regulatory, safety and ethical considerations. Vaccine 37:3267–3277. https://doi.org/10.1016/j.vaccine.2019.04.075

Kollie K, Siakeh A, Zawolo G, Wickenden A, Theobald S, Rogers E et al (2020) Donor reliance and the impact on neglected tropical disease programme delivery: reflections and solutions for change from programme management perspectives. Int Health 0:1–3. https://doi.org/10.1093/inthealth/ihaa090

Kung NY, Field HE, McLaughlin A, Edson D, Taylor M (2015) Flying-foxes in the Australian urban environment-community attitudes and opinions. One Health 1:24–30. https://doi.org/10.1016/j.onehlt.2015.07.002

La Vincente S, Kearns T, Connors C, Cameron S, Carapetis J, Andrews R (2009) Community management of endemic scabies in remote aboriginal communities of northern Australia: low treatment uptake and high ongoing acquisition. PLoS Negl Trop Dis 3(5):e444. https://doi.org/10.1371/journal.pntd.0000444

Labrique AB, Wadhwani C, Williams KA, Lamptey P, Hesp C, Luk R, Aerts A (2018) Best practices in scaling digital health in low and middle income countries. Glob Health 14(1):103. https://doi.org/10.1186/s12992-018-0424-z

Lamorde M, Mpimbaza A, Walwema R, Kamya M, Kapisi J, Kajumbula H et al (2018) A cross-cutting approach to surveillance and laboratory capacity as a platform to improve health security in Uganda. Health Secur 16(S1):S76–S86. https://doi.org/10.1089/hs.2018.0051

Lau LL, Hung N, Wilson K (2020) COVID-19 response strategies: considering inequalities between and within countries. Int J Equity Health 19:137. https://doi.org/10.1186/s12939-020-01254-9

Lee VJ, Aguilera X, Heymann D, Wilder-Smith A, Lee VJ, Heymann DL et al (2020) Preparedness for emerging epidemic threats: a lancet infectious diseases commission. Lancet Infect Dis 20(1):17–19. https://doi.org/10.1016/S1473-3099(19)30674-7

Machado FCG, Ferron MM, Barddal MT d M, Nascimento LA, Rosalen J, Avelino-Silva VI (2022) COVID-19 vaccination, incidence, and mortality rates among indigenous populations compared to the general population in Brazil: describing trends over time. Lancet Reg Health Am 13:100319. https://doi.org/10.1016/J.LANA.2022.100319

Maman I, Tchacondo T, Kere AB, Piten E, Beissner M, Kobara Y et al (2018) Risk factors for Mycobacterium ulcerans infection (Buruli ulcer) in Togo – a case-control study in zio and yoto districts of the mari-

time region. BMC Infect Dis 18(1):48. https://doi.org/10.1186/s12879-018-2958-3
McClarty L, Lazarus L, Pavlova D, Reza-Paul S, Balakireva O, Kimani J et al (2022) Socioeconomic burdens of the COVID-19 pandemic on LMIC populations with increased HIV vulnerabilities. Curr HIV/AIDS Rep 19:76–85. https://doi.org/10.1007/s11904-021-00591-w
McCollum AM, Damon IK (2014) Human monkeypox. Clin Infect Dis 58(2):260–267. https://doi.org/10.1093/cid/cit703
Mediannikov O, Diatta G, Fenollar F, Sokhna C, Trape J-F, Raoult D (n.d.) Tick-borne rickettsioses, neglected emerging diseases in rural senegal. PLoS Negl Trop Dis 4(9):e821. https://doi.org/10.1371/journal.pntd.0000821
Mishra V, Seyedzenouzi G, Almohtadi A, Chowdhury T, Khashkhusha A, Axiaq A et al (2021) Health inequalities during COVID-19 and their effects on morbidity and mortality. J Healthc Leadersh 13:19. https://doi.org/10.2147/JHL.S270175
Mohammed KA, Deb RM, Stanton MC, Molyneux DH (2012) Soil transmitted helminths and scabies in Zanzibar, Tanzania following mass drug administration for lymphatic filariasis – a rapid assessment methodology to assess impact. Parasit Vectors 5(1):299. https://doi.org/10.1186/1756-3305-5-299
Molla YB, Tomczyk S, Amberbir T, Tamiru A, Davey G (2012) Patients' perceptions of podoconiosis causes, prevention and consequences in East and West Gojam, Northern Ethiopia. BMC Public Health 12:828. https://doi.org/10.1186/1471-2458-12-828
Monroe AC, Ricketts TC, Savitz LA (1992) Cancer in rural versus urban populations: a review. J Rural Health 8:212–220. https://doi.org/10.1111/j.1748-0361.1992.tb00354.x
Moto JN, Maingi JM, Nyamache AK (2015) Prevalence of Tinea capitis in school going children from Mathare, informal settlement in Nairobi. BMC Res Notes 8:274. https://doi.org/10.1186/s13104-015-1240-7
Msyamboza KP, Mawaya LR, Kubwalo HW, Ng'oma D, Liabunya M, Manjolo S et al (2012) Burden of leprosy in Malawi: community camp-based cross-sectional study. BMC Int Health Hum Rights 12(1):12. https://doi.org/10.1186/1472-698X-12-12
Mubarak MY, Wagner AL, Asami M, Carlson BF, Boulton ML (2016) Hygienic practices and diarrheal illness among persons living in at-risk settings in Kabul, Afghanistan: a cross-sectional study. BMC Infect Dis 16(1):459. https://doi.org/10.1186/s12879-016-1789-3
Mukherjee S (2020) Disparities, desperation, and divisiveness: coping with COVID-19 in India. Psychol Trauma Theory Res Pract Policy 12(6):582. https://doi.org/10.1037/TRA0000682
Mushtaq A, Mehmood S, Rehman MAU, Younas A, Rehman MSU, Malik MF, Hyder MZ (2015) Polio in Pakistan: social constraints and travel implications. Travel Med Infect Dis 13:360–366. https://doi.org/10.1016/j.tmaid.2015.06.004
Neubeck L, Redfern JU, Fernandez R, Briffa T, Bauman A, Freedman SB (2009) Telehealth interventions for the secondary prevention of coronary heart disease: a systematic review. Eur J Prev Cardiol 6(3):281–289. https://doi.org/10.1097/HJR.0b013e32832a4e7a
Njoh AJ (2002) Barriers to community participation in development planning: lessons from the Mutengene (Cameroon) self-help water project. Community Dev J 37(3):233–248. https://doi.org/10.1093/cdj/37.3.233
Nkengasong J, Djoudalbaye B, Maiyegun O (2017) A new public health order for Africa's health security. Lancet Glob Health 5:e1064–e1065. https://doi.org/10.1016/S2214-109X(17)30363-7
Nuwematsiko R, Nabiryo M, Bomboka JB, Nalinya S, Musoke D, Okello D, Wanyenze RK (2022) Unintended socio-economic and health consequences of COVID-19 among slum dwellers in Kampala, Uganda. BMC Public Health 22(1):1–13. https://doi.org/10.1186/S12889-021-12453-6/FIGURES/5
OECD. (2022). First Lessons from government evaluations of COVID-19 responses: a synthesis. https://read.oecd-ilibrary.org/view/?ref=1125_1125436-7j5hea8nk4&title=First-lessons-from-government-evaluations-of-COVID-19-responses
Osadebe L, Hughes CM, Shongo Lushima R, Kabamba J, Nguete B, Malekani J et al (2017) Enhancing case definitions for surveillance of human monkeypox in the Democratic Republic of Congo. PLoS Negl Trop Dis 11(9):e0005857. https://doi.org/10.1371/journal.pntd.0005857
Palagyi A, Dodd R, Jan S, Nambiar D, Joshi R, Tian M et al (2019) Organisation of primary health care in the Asia-Pacific region: developing a prioritised research agenda analysis. BMJ Glob Health 4:1467. https://doi.org/10.1136/bmjgh-2019-001467
Peer N (2015) The converging burdens of infectious and non-communicable diseases in rural-to-urban migrant Sub-Saharan African populations: a focus on HIV/AIDS, tuberculosis and cardio-metabolic diseases. Trop Dis Travel Med Vaccines 1:6. https://doi.org/10.1186/s40794-015-0007-4
Pigott DM, Deshpande A, Letourneau I, Morozoff C, Reiner RCJ, Kraemer MUG et al (2017) Local, national, and regional viral haemorrhagic fever pandemic potential in Africa: a multistage analysis. Lancet 390(10113):2662–2672. https://doi.org/10.1016/S0140-6736(17)32092-5
Priorities P (2022) Emerging risks to the recovery. World development report 2022: finance for an equitable recovery, 49–73. doi:https://doi.org/10.1596/978-1-4648-1730-4_ch1
Rahman F (2020) The Malaysian response to COVID-19: building preparedness for 'surge capacity', testing efficiency, and containment – from the desk of the director-general of health Malaysia. DG of health Malaysia blog, (April 2020), 1–22
Rampal L, Seng LB, Choolani M, Ganasegeran K, Pramanick A, Vallibhakara SAO et al (2020) Battling covid-19 pandemic waves in six south-east asian

countries: a real-time consensus review. Med J Malays 75(6):613–625

Rao B, Lombardi A (2009) Telemedicine: current status in developed and developing countries. J Drugs Dermatol 8(4):371–375

Rimi NA, Sultana R, Ishtiak-Ahmed K, Rahman MZ, Hasin M, Islam MS et al (2016) Understanding the failure of a behavior change intervention to reduce risk behaviors for avian influenza transmission among backyard poultry raisers in rural Bangladesh: a focused ethnography. BMC Public Health 16(1):858. https://doi.org/10.1186/s12889-016-3543-6

Rosa WE, Ahmed E, Chaila MJ, Chansa A, Cordoba MA, Dowla R et al (2022) Can you hear us now? Equity in global advocacy for palliative care. J Pain Symptom Manag 64(4):e217–e226. https://doi.org/10.1016/j.jpainsymman.2022.07.004

Ryll R, Eiden M, Heuser E, Weinhardt M, Ziege M, Höper D et al (2018) Hepatitis E virus in feral rabbits along a rural-urban transect in central Germany. Infect Genet Evol 61:155–159. https://doi.org/10.1016/j.meegid.2018.03.019

Sabes-Figuera R, McCrone P, Bogic M, Ajdukovic D, Franciskovic T, Colombini N et al (2012) Long-term impact of war on healthcare costs: an eight-country study. PLoS One 7(1):e29603. https://doi.org/10.1371/journal.pone.0029603

Sarfraz Z, Sarfraz A, Jaiswal V, Poudel S, Bano S, Hanif M et al (2022) The past, present and future of COVID-19 associated Mucormycosis: a rapid review. J Prim Care Community Health 13:21501319221099476. https://doi.org/10.1177/21501319221099476

Schuster A, Lesshafft H, Talhari S, Guedes de Oliveira S, Ignatius R, Feldmeier H (2011) Life quality impairment caused by hookworm-related cutaneous larva migrans in resource-poor communities in Manaus, Brazil. PLoS Negl Trop Dis 5(11):e1355. https://doi.org/10.1371/journal.pntd.0001355

Seimenis A (2012) Zoonoses and poverty – a long road to the alleviation of suffering. Vet Ital 48(1):5–13

Sharma A, Badea M, Tiwari S, Marty JL (2021) Wearable biosensors: an alternative and practical approach in healthcare and disease monitoring. Molecules 26:748. https://doi.org/10.3390/molecules26030748

Smith KB, Humphreys JS, Wilson MG, a. (2008) Addressing the health disadvantage of rural populations: how does epidemiological evidence inform rural health policies and research? Aust J Rural Health 16(2):56–66. https://doi.org/10.1111/j.1440-1584.2008.00953.x

Snyder RE, Marlow MA, Phuphanich ME, Riley LW, Maciel ELN (2016) Risk factors for differential outcome following directly observed treatment (DOT) of slum and non-slum tuberculosis patients: a retrospective cohort study. BMC Infect Dis 16(1):1–7. https://doi.org/10.1186/s12879-016-1835-1

SteelFisher GK, Blendon RJ, Guirguis S, Brulé A, Lasala-Blanco N, Coleman M et al (2015) Threats to polio eradication in high-conflict areas in Pakistan and Nigeria: a polling study of caregivers of children younger than 5 years. Lancet Infect Dis 15(10):1183–1192. https://doi.org/10.1016/S1473-3099(15)00178-4

Stern D, Olson VA, Smith SK, Pietraszczyk M, Miller L, Miethe P et al (2016) Rapid and sensitive point-of-care detection of Orthopoxviruses by ABICAP immunofiltration. Virol J 13(1):207. https://doi.org/10.1186/s12985-016-0665-5

Sutton TC (2018) The pandemic threat of emerging H5 and H7 avian influenza viruses. Viruses 10(9):461. https://doi.org/10.3390/v10090461

Tagoe ET, Sheikh N, Morton A, Nonvignon J, Sarker AR, Williams L, Megiddo I (2021) COVID-19 vaccination in lower-middle income countries: national stakeholder views on challenges, barriers, and potential solutions. Front Public Health 9:709127. https://doi.org/10.3389/fpubh.2021.709127

Tasnim S, Hossain M, Mazumder H (2020) Impact of rumors and misinformation on COVID-19 in social media. J Prev Med Public Health 53:171–174. https://doi.org/10.3961/JPMPH.20.094

Terefe M, Solomon C, Menza K, Bedemo A (2017) Determinants of urban poverty: the case of Nekemte town, eastern Wollega zone of Oromia regional state. J Poverty Dev Invest 32

Tran TPT, Le TH, Nguyen TNP, Hoang VM (2020) Rapid response to the Covid-19 pandemic: Vietnam government's experience and preliminary success. J Glob Health 10(2):1–14. https://doi.org/10.7189/jogh.10.020502

Troiano G, Nardi A (2020) Vaccine hesitancy in the era of COVID-19. Public Health 194:245–251

Tusabe F, Tahir IM, Akpa CI, Mtaki V, Baryamujura J, Kamau B et al (2022) Lessons learned from the Ebola virus disease and COVID-19 preparedness to respond to the human Monkeypox virus outbreak in low- and middle-income countries. Infect Drug Resist 15:6279–6286. https://doi.org/10.2147/IDR.S384348

UNAIDS. (2016) 2016 United nations political declaration on ending AIDS sets world on the fast-track to end the epidemic by 2030. http://www.unaids.org/sites/default/files/20160608_PS_HLM_Political_Declaration_final.pdf

Veldhuis CB, Nesoff ED, McKowen ALW, Rice DR, Ghoneima H, Wootton AR et al (2021) Addressing the critical need for long-term mental health data during the COVID-19 pandemic: changes in mental health from April to September 2020. Prev Med 146:106465. https://doi.org/10.1016/j.ypmed.2021.106465

WHO. (2015) WHO. Global technical strategy for malaria 2016–2030. Geneva: World Health Organization (WHO); 2015 (http://www.who.int/malaria/areas/global_technical_strategy/en, Accessed 19 Oct 2017). World Health Organization. https://doi.org/ISBN: 978 92 4 156499 1

WHO Health Workforce Department. (2021) The impact of COVID-19 on health and care workers: a closer look at deaths. (September)

Wichmann B, Wichmann R (2022) COVID-19 and indigenous health in the Brazilian Amazon. Econ Model 115:105962. https://doi.org/10.1016/j.econmod.2022.105962

Wignjadiputro I, Widaningrum C, Setiawaty V, Widuri Wulandari E, Sihombing S, Prasetyo WA et al (2020) Whole–of–society approach for influenza pandemic epicenter containment exercise in Indonesia. J Infect Public Health 13(7):994–997. https://doi.org/10.1016/j.jiph.2019.12.009

Wijesinghe PR, Ofrin RH, Bhola AK, Inbanathan FY, Bezbaruah S (2020) Pandemic influenza preparedness in the WHO South-East Asia region: a model for planning regional preparedness for other priority high-threat pathogens. WHO South East Asia J Public Health 9(1):43–49. https://doi.org/10.4103/2224-3151.282995

World Health Organization. (2020a) Global tuberculosis report 2020. https://apps.who.int/iris/bitstream/handle/10665/336069/9789240013131-eng.pdf

World Health Organization. (2020b) Prioritizing diseases for research and development in emergency contexts. World-Health-Organization. https://www.who.int/activities/prioritizing-diseases-for-research-and-development-in-emergency-contexts

Wu X, Oliveria SA, Yagerman S, Chen L, Defazio J, Braun R, Marghoob AA (2015) Feasibility and efficacy of patient-initiated mobile teledermoscopy for short-term monitoring of clinically atypical nevi. JAMA Dermatol 151(5):489–496. https://doi.org/10.1001/jamadermatol.2014.3837

Zafar SN, McQueen KAK (2011) Surgery, public health, and Pakistan. World J Surg 35(12):2625–2634. https://doi.org/10.1007/s00268-011-1304-3

Zhang Y, Yu Y-S, Tang Z-H, Chen X-H, Zang G-Q (2014) Crack down on medical corruption: an urgent matter in China. Eur J Intern Med 25(1):e2–e3. https://doi.org/10.1016/j.ejim.2013.10.012

Zhang J, Li J, Kwak D-H, Zolotarev OV, Larnyo E, Dai B et al (2022) Impact of actual use behavior of healthcare wearable devices on quality of life: a cross-sectional survey of people with dementia and their caregivers in Ghana. Healthcare 10(2):275. https://doi.org/10.3390/healthcare10020275

Ziegler T, Mamahit A, Cox NJ (2018) 65 years of influenza surveillance by a World Health Organization-coordinated global network. Influenza Other Respir Viruses 12(5):558–565. https://doi.org/10.1111/irv.12570

Index

F. A. Mphande, *Sustainable Health in Low and Middle Income Countries*, Sustainable Development Goals Series, https://doi.org/10.1007/978-981-99-4254-1

The manufacturer's authorised representative in the EU is Springer Nature Customer Service Centre GmbH, Europaplatz 3, 69115 Heidelberg, Germany. If you have any concerns regarding our products, please contact ProductSafety@springernature.com

Printed and bound by CPI Group (UK) Ltd, Croydon, CR0 4YY
01/10/2026
02225877-0019